Electron-Molecule Scattering

WILEY SERIES IN PLASMA PHYSICS

SANBORN C. BROWN ADVISORY EDITOR
RESEARCH LABORATORY OF ELECTRONICS
MASSACHUSETTS INSTITUTE OF TECHNOLOGY

BROWN · ELECTRON-MOLECULE SCATTERING
GOLANT, ZHILINSKY, AND SAKHAROV · FUNDAMENTALS OF PLASMA PHYSICS
HUXLEY AND CROMPTON · THE DIFFUSION AND DRIFT OF ELECTRONS IN GASES
MITCHNER AND KRUGER · PARTIALLY IONIZED GASES
McDANIEL AND MASON · THE MOBILITY AND DIFFUSION OF IONS IN GASES
GILARDINI · LOW ENERGY ELECTRON COLLISIONS IN GASES
TIDMAN AND KRALL · SHOCK WAVES IN COLLISIONLESS PLASMAS
NASSER · FUNDAMENTALS OF GASEOUS IONIZATION IN PLASMA ELECTRONICS
LICHTENBERG · PHASE-SPACE DYNAMICS OF PARTICLES
BROWN · INTRODUCTION TO ELECTRICAL DISCHARGES IN GASES
BEKEFI · RADIATION PROCESSES IN PLASMAS
MACDONALD · MICROWAVE BREAKDOWN IN GASES
HEALD AND WHARTON · PLASMA DIAGNOSTICS WITH MICROWAVES
MCDANIEL · COLLISION PHENOMENA IN IONIZED GASES

ELECTRON-MOLECULE SCATTERING

Edited by
SANBORN C. BROWN

Professor of Physics Emeritus
Massachusetts Institute of Technology

A WILEY INTERSCIENCE PUBLICATION

JOHN WILEY & SONS, New York · Chichester · Brisbane · Toronto

Library of Congress Cataloging in Publication Data:

Main entry under title:
Electron-molecule scattering.

(Wiley series in plasma physics)
"A Wiley Interscience publication."
In memory of George J. Schulz.
Includes index.
1. Electron-molecule scattering. 2. Schulz, George J., 1925– I. Brown, Sanborn Conner, 1913– II. Schulz, George J., 1925–

QC793.5.E628E39 539.7'211 79-12705
ISBN 0-471-05205-1

Printed in the United States of America

10 9 8 7 6 5 4 3 2 1

Scientific Contributors

George J. Schulz
Mason Laboratory
Yale University
New Haven, Connecticut

Manfred A. Biondi
Department of Physics and Astronomy
University of Pittsburgh
Pittsburgh, Pennsylvania

Arthur V. Phelps
Joint Institute for Laboratory Astrophysics
National Bureau of Standards and University of Colorado
Boulder, Colorado

Franz Linder
Department of Physics
University of Kaiserslautern
675Kaiserslautern, Federal Republic of Germany

Neal Lane
Physics Department
Rice University
Houston, Texas

Harrie S. W. Massey
Department of Physics and Astronomy
University College London
London, England

This original portrait of George J. Schulz, 1925–1976, was contributed by Patrice Donnell.

Tributes to George J. Schulz

About eight or nine years ago, George Schulz took me around his laboratory here at Yale, and to everyone he said: "This is my old professor." So as his old professor, I take it as a great honor to open this memorial to George Schulz. A teacher always has a special feeling for his graduate students, and when they do especially well, he is especially proud. Surely this is my feeling about George.

One of George's characteristics which infected us all was his enthusiasm. For those of you who got to know him only for the last 10 or 15 years, you should have known him when he was in his twenties. With him in our research group, it was an exciting place and he had a large effect on many of the other graduate students at that time: Larry Gould, Hutch Looney, Karl Persson, Art Phelps, and Harry Dreicer. They were all most affected by his optimism and his rare good humor, particularly when things were difficult. It was a wonderful thing to watch. Although this is a sad occasion because we have to honor the memory of George Schulz, it is with a feeling of real satisfaction that we remember him as he would have wanted to be remembered—as a physicist.

SANBORN C. BROWN

It is a great honor for me to be asked to contribute to this memorial for George J. Schulz. George Schulz appreciated theory and recognized its important, if somewhat distinct, roles of prediction and interpretation. He was most perceptive, and demanded that the theorist keep these distinctions in mind and describe his work accordingly. He very much enjoyed the game "What would you expect to happen if . . .," holding back a few moments on the results of a measurement just made. If your response agreed with his data he was pleased because that meant "Theory was doing well." If you got it wrong, he was also happy because "The theorist had learned something in the experience." George had a way of reminding you that you were doing science and science is a rather rigorous activity. George Schulz had tremendous impact on theoretical developments in electron scattering and significant progress has been made.

NEAL LANE

I knew George Schulz quite well. He always opened a conversation with me after resonances were discovered by pointing out that these resonances nearly always lie close to and just below an excited state. "Why shouldn't they be anywhere else?" He asked me that question every time I met him, so you can judge that I never satisfied him on any one occasion.

It always seemed to me that the beauty of the experiments he carried out is hard to express in words. There is no doubt at all that he was one of those exceptional experimenters who seem to have "green fingers" in producing definite results no matter how difficult the subject of research may be. His group devised the most suitable experiments, carried them out with meticulous care, and presented their results in a definite, clear-cut, and relevant way. I think that George Schulz was an absolute master in these respects.

HARRIE S. W. MASSEY

You will find in this volume many works that are based on results by George Schulz, and even direct quotations from his works, mostly done at Yale. For example you will see several double electrostatic analysers here, together with vacuum systems, molecular beam sources, electron multipliers, computers, and plotters that can be checked out to graduate or post doctoral students as they come, for George was a great teacher. Clearly the action passed from the laboratory to the lunch table where the possibility and usefulness of new measurements were argued and the observed wiggles, good or bad, were discussed. This was George's method of graduate teaching and of research, and we know with what enthusiasm he pursued it. He carried it over to conferences and it was he who introduced the "workshop" in the Gaseous Electronics Conferences, where it has been a great success.

Schulz had a superb sense of what was possible experimentally and worthwhile when done, but he needed help. First, he needed a theorist to interpret his measurements and at Yale this was Herzenberg. I believe that the meeting of these two minds, one knowing the wave functions and the other helping the first to put them graphically on paper, was tremendously productive. The second necessity was a competent and cooperative machine shop, such as the one that thrived at Yale. George often lunched with the machinists so he knew them well. He was proud of their craftsmanship.

George was ardently American to a degree that would have been chauvinist had he been born here, but in fact expressed a desire to integrate into the society that had received him. He did not want to flaunt ethnic, religious, or social differences, and I can remember his telling me how happy he was one day to see a workman drive up to work at Pennsylvania State University in a Cadillac. His loud laugh and bold manner expressed a feeling of comfort with his surroundings and of adaptation to this country, but it also protected an internal reserve and sensitivity. He never spoke of his life before coming here and his friends know little of it. We know, of course, that he was born in Brno, in Moravia in 1925, and that he worked in a German labor camp during the war. Hence George had known only economic depression, war, invasion and slavery, and he had seen the invaders use the rivalries among Czechs, Hungarians, Slavs, Poles, and Jews to their advantage. It is difficult to imagine George's independent spirit as downtrodden so that the war period must have been extremely painful to him. I feel that this was a memory George would not want to have recalled.

The postwar period brought two waves of brilliant students to M.I.T. First the demobilized military or war workers which included Biondi, Rose, and Phelps; then the foreign students: MacDonald, Schulz, Persson, and Buchsbaum. George started working with Nottingham but shifted to Sandy Brown's group and undertook a thesis comparing microwave to probe measurements. He then went to Westinghouse to work in Dan Alpert's group. Schulz designed, with Phelps, an ionization gauge that worked at higher pressures and worked with Chantry on dissociative attachment. He and Fox measured the cross section for excitation of helium to the metastable levels using the differential retarding potential method to determine the electron energy to within $\frac{1}{2}$ volt! This work was the root from which all his later work grew, for he had the brilliant idea of putting a slight retarding potential after the collision chamber as well as before, thus creating a slow electron trap. This seemed like a trivial modification and there are today many electron and ion traps, but Schulz saw that this device was a sensitive instrument with which he could detect the onset of any inelastic electron collision, measure the slope of the cross section, and detect any resonances along it; and he did.

The shape of resonances of N_2, H_2, O_2, CO, and so on all followed. Although this method was then 10^4 times more sensitive than deflection methods, it was not as well suited to detect atomic resonances in which no energy is lost, so the Feshbach resonance in helium was found with a primitive differential electrostatic analyser. The discovery of these resonances won Schulz the Davisson–Germer prize.

New resonances were accumulating fast and though their nature was not evident, Schulz interpreted the molecular resonances as being temporary negative ions and therefore exhibiting the vibrational spectrum of molecular ions. George had now found his field, resonances. He was making new discoveries in this field, and these discoveries were falling into recognizable patterns. Such is the great joy of research.

William P. Allis

Preface

When George Schulz took his Ph.D. in 1954, the field of electron-atom and electron-molecule collisions stood where the pioneers had left it 20 years before. A shroud of inadequate experimental resolution hid most of the striking features of the collisions. George Schulz was in the vanguard of those who broke through to resolutions better than 100 meV, and the first to give us a glimpse of the fascinating world of resonances which lay beyond. Following his own dictum, "First do what you can, then what you want," he started by uncovering the famous resonances in nitrogen and helium, which, 15 years later, are still the workhorses of the theorists. Where George Schulz led, others followed; his discoveries ushered in a renaissance in electron spectroscopy.

Nowadays, George Schulz's union of the excitement of discovery with the mastery of experimental precision is commonplace in many laboratories. Experimenters the world over think of him as the father who led them into the field. It was therefore a fitting tribute to George Schulz's memory to set up a George J. Schulz Lecture at Yale University where he had spent so much of his professional life. The first Memorial Lecture was given by Professor Manfred A. Biondi on October 28, 1977. The following day was occupied with a number of contributions to a symposium on electron-molecule scattering. These lectures formed an excellent summary of the problems in this field and the direction in which the field is going. It was felt that their publication in book form would be useful in getting graduate students started in the study of electron-molecule scattering.

Since the focus of the Memorial Lecture and the Symposium was the work which George Schulz himself had done, it seems proper to include in this collection of papers "A Review of Vibrational Excitations of Molecules by Electron Impact at Low Energies" published in *Principles of Laser Plasmas*, edited by George Bekefi, a review completed just before George Schulz's untimely death.

The lecture and symposium were made possible by contributions from many of George Schulz's friends and colleagues and from Yale University, the Yale Science and Engineering Association, the Yale chapter of Sigma Xi, the Westinghouse Electric Corporation, and the Yale Department of Engineering and Applied Science. They were arranged and coordinated by Professor Arvid Herzenberg.

SANBORN C. BROWN

Hemlock Corner
Henniker, NH
May 1979

Contents

Electron-Molecule Scattering

CHAPTER 1

A Review of Vibrational Excitations of Molecules by Electron Impact at Low Energies

GEORGE J. SCHULZ
Mason Laboratory
Yale University
New Haven, Connecticut

Reprinted with permission from *Principles of Laser Plasmas*, George Bekefi, editor. John Wiley and Sons, New York (1976) Chapter 2.

1 INTRODUCTION

This chapter deals with one of the primary electron impact processes that has a bearing on gas lasers, namely vibrational excitation by electron impact. A large group of gas lasers relies on electron impact phenomena, and subsequent reactions, to achieve population inversion. But other classes of gas lasers, akin to solid-state and liquid-state lasers, produce the population inversion by optical pumping or chemical reactions, and for these lasers electron collisions are not of primary interest. Nevertheless it appears that the unique feature of gases, that population inversion can be achieved through electron impact and subsequent excitation transfer, presents unique advantages in terms of efficiency and simplicity. The simplest arrangement for achieving population inversion in a gas, at least from a conceptual viewpoint, would involve the use of a "monoenergetic" beam of electrons that would excite preferentially the upper laser state. In such an arrangement it would be required that the lower laser state have a cross section considerably smaller than the cross section of the upper laser state or that the lower laser state could be depopulated by other means. Despite the attractiveness and the simplicity of this scheme it has been achieved only rarely (Tien, MacNair, and Hodges, 1964). Instead the preferred mode of population involves a gas discharge, usually in a mixture of gases. This scheme has been pioneered by Javan, Bennett, and Herriott (1961) for atomic systems and by Patel (1964, 1965) for molecular gases.

The need to understand population inversion and laser action in gas discharges in terms of electron impact on molecules and subsequent reactions puts enormous demands on our knowledge of the pertinent cross sections. Ideally we would need to know a whole *set* of cross sections for elastic and inelastic electron collisions in the energy range from 0 to about 20 eV for all the molecules or atoms of the gas mixture. In addition, we would need to know the cross sections for excited molecular species (vibrationally and electronically excited states) interacting with the other molecular species. Negative and positive ions formed by electron impact also participate in reactions, forming new ionic and neutral species. If all these cross sections were reliably known, one could calculate the electron energy distribution functions, inversion, and instabilities and generally characterize the medium accurately. Unfortunately many of the cross sections for molecules that have lasing properties cannot yet be supplied and we must be satisfied with a rather fragmentary knowledge, although our understanding of cross sections has made dramatic advances in the past decade. These advances turn out to be particularly important in the range of low energies (0 to 20 eV).

It is not appropriate in these pages to review the totality of cross sections that play a role in lasers. Excellent reviews on many of these are available

(Massey, 1969; McDaniel, 1964; Hasted, 1964; Brown, 1959; Phelps, 1968; Gilardini, 1972) and their applicability to lasers has also been reviewed (Patel, 1968). This chapter is limited to an overview of vibrational cross sections, in which vibrational energy is supplied to the target molecules by electrons. Stress is placed on the physical processes and the fundamental character of the particular inelastic processes.

Vibrational excitation actually plays an important role in the science of gas lasers. The most efficient laser systems designed to date, namely, the N_2–CO_2 and the CO lasers, rely on vibrational transitions for laser action. In the N_2–CO_2 system the large vibrational cross section, near 2.3 eV in N_2, resulting from a resonance, causes the electron distribution function to be sharply cut off. This results in an efficient population of vibrational levels of N_2, with subsequent transfer to the upper laser state of CO_2 (i.e., the asymmetric stretch mode). We note that no inversion of the population of vibrational levels occurs in the N_2 system. In fact, resonance processes, which often lead to large vibrational cross sections, do not generally lead to population inversion. In the CO infrared laser (wavelength near 5 to 6 μm), laser action takes place by transitions between high vibrational states of CO (e.g., $v = 10 \rightarrow 9$, $9 \rightarrow 8$, $8 \rightarrow 7$), generally in the *P*-branch. As in N_2, the high vibrational levels of CO are populated by the CO resonance, although other mechanisms may also play a role. Usually no inversion of vibrational levels takes place, but the rotational levels associated with the *P*-branch exhibit inversion.

Occasionally, complete inversion has been observed in pulsed CO discharges, as indicated by lasing in the *R*-branch (Jeffers and Wiswall, 1971). The complete inversion is attributed by Jeffers et al. to collisions between vibrationally excited species taking into account the anharmonic nature of the vibrational levels.

2 RESONANT AND NONRESONANT EXCITATIONS

It is desirable, for a physical understanding of the underlying mechanisms, to divide the discussion of cross sections into two broad categories, resonant and nonresonant processes (Chen, 1964). All inelastic processes can proceed by both these mechanisms, but often the vibrational, rotational, and electronic cross sections are affected in major ways by the presence of resonances. Resonances occurring in electron impact often enhance inelastic cross sections by orders of magnitude and sometimes the energy dependence of these cross sections exhibits oscillatory structure or isolated peaks. Nonresonant cross sections generally show a smooth, slowly varying energy dependence.

The division of cross sections into resonant and nonresonant regimes need not be a priori if a complete calculation is made. Instead by including centrifugal, polarization, static, and exchange effects in the potential, a complete calculation would automatically acount for all resonances, as well as for the nonresonant cross section. Such calculations, although feasible for atoms, have only been done for H_2 and are only being introduced at present for N_2 (Chandra and Temkin, 1976). Until the complete calculation becomes commonplace, a division of cross sections into resonant and nonresonant components appears to be most useful. Even when a complete calculation can be made, such a division elucidates the essential physics of the problem. Often different inelastic processes (e.g., rotational and vibrational excitation) can be accounted for by the decay of a single resonance, which is characterized by its lifetime (width), its energy, and its branching ratio. As will be seen, the resonant regimes are often most interesting for laser applications.

2.1 Resonant Vibrational Excitation

Resonances in electron scattering have been recently reviewed, both for atoms and for diatomic molecules (Schulz, 1973). These resonances occur in more or less well-defined energy ranges, and at these energies electrons spend much more time in the neighborhood of the molecule than is characteristic of the normal transit time. The electrons that are trapped in the neighborhood of the molecule have an increased interaction time and cause a more efficient distortion of the molecule. This distortion leads to an enhancement in vibrational excitation.

The nature of vibrational excitation via resonance processes is strongly influenced by the lifetime of the resonance. These resonances can lead to broad peaks (several electron volts wide) that do not exhibit structure or to peaks that do exhibit fine structure, or to bands of isolated spikes, or to ultrabroad peaks, which can be 10 eV wide. We discuss these cases under separate headings.

2.1.1 Broad Peaks without Fine Structure—the Impulse Limit. When the lifetime of the resonance is much smaller than a typical vibrational period ($\tau \ll 10^{-14}$ sec), the nuclei cannot vibrate during the lifetime of the resonance. The incident electron under these circumstances attaches itself to the molecule (forming a particular intermediate state) and the system decays by the emission of the extra electron, leaving the molecule in a vibrationally excited level. Because of the uncertainty principle, the intermediate state has a large width in energy ($\Gamma \cdot \tau \cong \hbar$, where Γ is the width of the state and τ its lifetime) and this leads to a *broad hump* in the energy dependence of the vibrational cross section. A characteristic of this category of resonances is the *rapidly diminishing ratio of the magnitudes of the vibrational excitation cross sections to successive vibrational levels.*

The impulse limit, which represents the short-lifetime limit of resonances, has been discussed by Herzenberg and Mandl (1962) and by Dubé and Herzenberg (1975). The principal physical cause for vibrational excitation in this limit is the distortion (amplitude modulation) of the nuclear wavepacket in the negative ion. The components of this distorted wavepacket when expanded in the vibrational wavefunctions of the neutral molecule exhibit an overlap with the excited vibrational states. The motion of the wavepacket itself, which is small but finite during the lifetime of the resonance, is not considered in the impulse limit, although in longer-lived resonances (boomerang states, for example) this motion is very important.

Examples of vibrational excitation through short-lived resonances are found in H_2 and N_2O near 2 eV and probably in N_2 near 8 and 14 eV.

Somewhat akin to the effect of these short-lived resonances are resonances whose potential energy curve is repulsive. Such resonances also lead to a broad hump, although the lifetime of the state may be somewhat longer (e.g., 10^{-14} sec).

2.1.2 Peaks with Shifting Fine Structure—Boomerang States. Once the lifetime of the compound state (i.e., the resonance lifetime) significantly exceeds the time required by the electron to travel directly across the molecular diameter and *becomes comparable to a typical vibrational period*, the nuclei can perform one (or a few) vibrations before the compound state has decayed. Such resonances have lifetimes of about 10^{-14} sec.

The behavior of the vibrational cross section under these circumstances has been successfully explained by the boomerang model proposed by Herzenberg (1968) for N_2 near 2.3 eV. The model proposes that once the electron attaches itself to the N_2 molecule, the nuclear wavefunction propagates across the N_2^- potential well in such a manner that only one reflection of the wavepacket is important. The wavepacket is completely annihilated by autodetachment when it reaches the end of the first cycle. Detailed calculations by Birtwistle and Herzenberg (1971) and comparison with experiment substantiate the model and confirm that only a single outgoing and a single reflected wave need to be considered. This process is indicated schematically in Fig. 1.1. The superposition of these two waves produces a standing wave whose nodes drift slowly in position as the incident electron energy is varied. Calculation of the overlap integral between this standing wave and the vibrational states of the neutral molecule reveals that the energy dependences of the *vibrational cross sections*

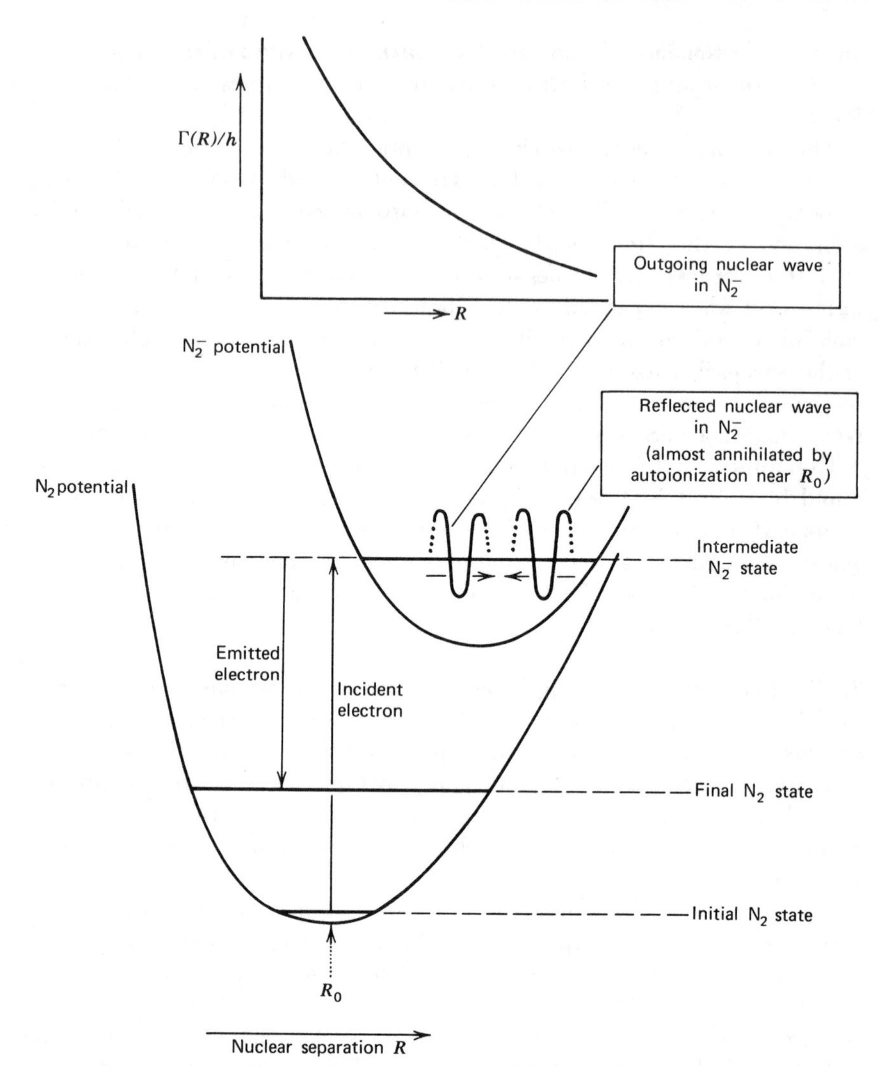

Fig. 1.1 The "boomerang" model of the nuclear wavefunction applied to the N_2^- ion. This model is discussed by Herzenberg (1968). It is based on the assumption that the magnitude and R-dependence of the width $\Gamma(R)$ are such that only a single outgoing and a single reflected wave are important. (From Birtwistle and Herzenberg, 1971.)

oscillate and that the *peaks of these oscillations occur at different energies*, depending on the final channels of excitation. The functional dependence of the autoionization width $\Gamma(R)$ on internuclear separation R is governed by the relative locations of the N_2 and N_2^- potential-energy curves and by the centrifugal barrier, which traps the electron. The variation of $\Gamma(R)$ is schematically indicated at the top of Fig. 1.1. Finally in order to reproduce qualitative agreement with the experimental data, the magnitude of $\Gamma(R)$ has to be adjusted to provide the desired annihilation rate for the nuclear wavefunction. An essential feature of this model is the interference between the outgoing and reflected nuclear wavepackets, without which it is impossible to reproduce the oscillatory energy dependence of the cross sections combined with a shift in the peaks.

2.1.3 Sharp Spikes—Long-lived Resonances. When the lifetime of the compound state increases further, many vibrations of the nuclei can take place during the lifetime of the compound state. This leads to well-developed vibrational levels of the compound state. Decay of these vibrational levels of the compound state takes place to various vibrational levels of the neutral molecule by the emission of the extra electron. *The vibrational cross sections thus exhibit sharp spikes at the energies of the vibrational levels of the compound state* (Herzenberg and Mandl, 1962).

An example of a system that falls into this category is the oxygen molecule (see Sec. 5). Two isolated spikes in the range 11.4 to 12 eV in N_2 also fall into this category. Here we are dealing with the lowest Feshbach-type resonance of the N_2 system. This lowest Feshbach resonance can be visualized with the help of Fig. 1.2. We can imagine that it is formed by the addition of two

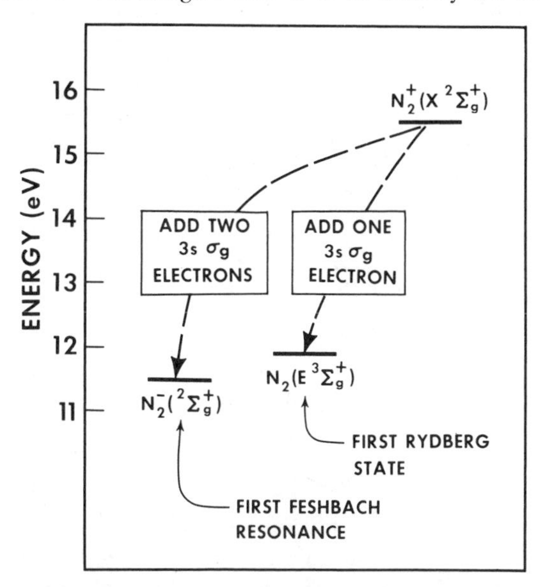

Fig. 1.2 Genealogy of the first Feshbach resonance in N_2. The diagram shows schematically that the $^2\Sigma_g^+$ resonance can be pictured as being derived from the ground state of the positive ion by the addition of two $3s\sigma_g$ electrons.

electrons (in $3s\sigma_g$ orbitals) to the ground state of the positive ion ($X^2\Sigma_g^+$). The resulting first Feshbach resonance, $^2\Sigma_g^+$ lies at 11.48 eV. It is relatively long-lived with respect to decay into the ground electronic state of the neutral molecule because its core configuration resembles the positive ion rather than the ground electronic state of the neutral and thus considerable rearrangement is needed for the decay.

2.1.4 Ultrabroad Peaks—Multiply Excited States. Peaks in the vibrational cross section with a width of about 10 eV have been recently observed and it is highly improbable that peaks of such width could result from a single resonance. Instead Pavlovic et al. (1972), who first observed this effect in N_2, interpreted it in terms of a large number of closely spaced resonances, which are unresolved and overlapping. The resonances involved are principally of the type formed by the addition of an electron to *doubly excited* valence states of the neutral molecule. It is estimated that the

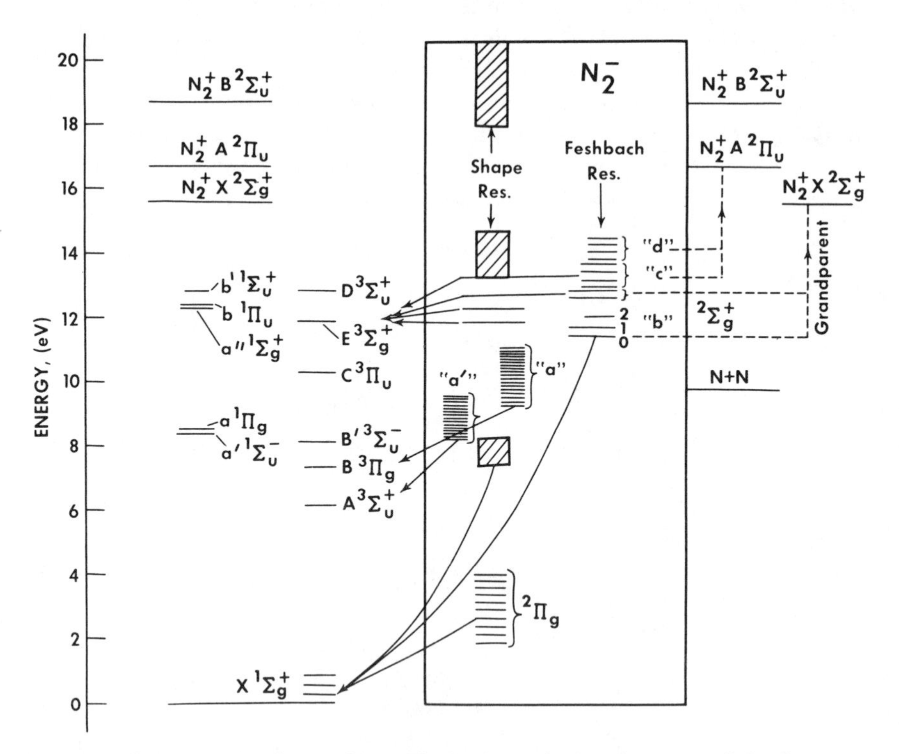

Fig. 1.3 Energy levels of N_2 and N_2^-. The levels in the boxed portion of the figure are resonances, divided into shape resonances and Feshbach resonances. A few decay channels are indicated. The dashed lines leading to positive-ion states indicate the grandparents of Feshbach resonances.

average spacing of such resonances in the 22 eV region of N_2 is about 0.05 eV.

2.1.5 Global View of Resonant Vibrational Excitation: N_2 in the Energy Range 1 to 30 eV. As an example of the preceding general discussion, we now discuss the features of the vibrational cross section in N_2. The nitrogen molecule serves as a good vehicle for such a discussion because it exhibits essentially all the features discussed earlier. Because N_2 does not have an electric dipole moment, the direct component of vibrational excitation is small and the resonance component clearly dominates.

Figure 1.3 shows, in the boxed position of the figure, the resonances in N_2 as they are known to date. All these resonance energies are deduced from experiments, although a priori calculations of some of these positions are now also feasible (Krauss and Mies, 1970). The bulk of the vibrational cross sections proceeds via one or the other of the resonances shown in Fig. 1.3,

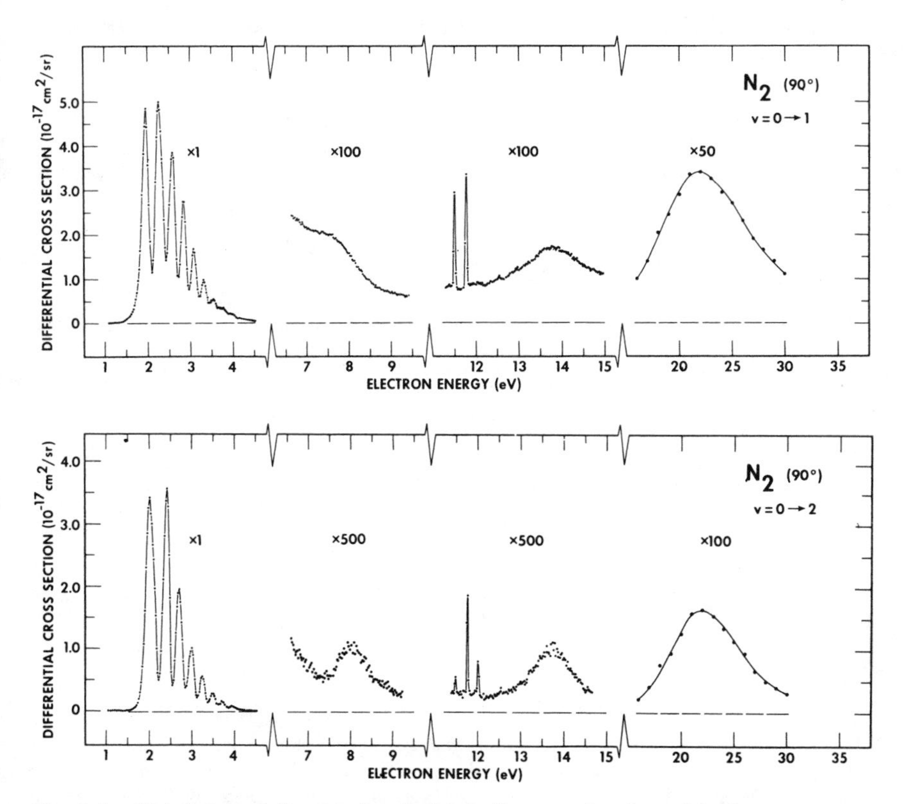

Fig. 1.4 Global view of vibrational excitation in N_2 to $v = 1$ and $v = 2$ in the energy range 1 − 30 eV. (From Wong and Schulz, unpublished.)

but not all resonances cause a measurable vibrational excitation of the ground electronic state. This can be seen from Fig. 1.4, which shows the differential vibrational cross section $v=1$ and $v=2$ for electrons in the energy range 1 to 30 eV. This curve is obtained with an electron spectrometer (Fig. 1.5) in which electrons with a small energy spread (~35 meV) are crossed with a beam of molecules and the electrons scattered at an angle of 90° (referred to the incident electron beam) are energy-analyzed. A peak in the energy loss spectrum, characteristic of the vibrational transition in N_2, is associated with vibrational excitation of N_2.

Figure 1.4 can be interpreted on the basis of at least five resonance regions located in the energy ranges 1 to 4 eV, 7 to 8 eV, 11 to 12 eV, 13 to 14 eV, and 15 to 30 eV. The characteristics of the resonances (i.e., their lifetimes and potential energy curves) in each of these energy ranges are different and this influences the features observed. The region from 1 to 4 eV, which is later discussed in more detail, results from the lowest shape resonance, which in the case of N_2 exhibits an intermediate lifetime,

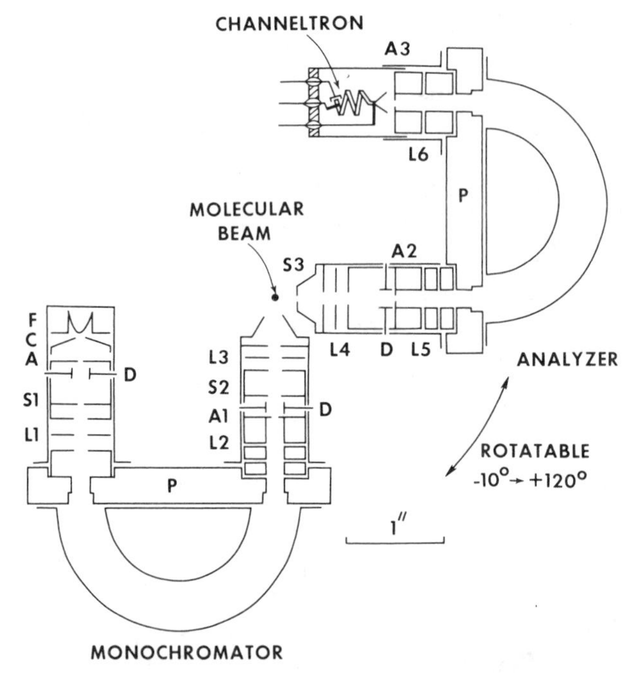

Fig. 1.5 Schematic diagram of the double hemispherical electron spectrometer. Lenses are designated by *L*, angular stops by *S*, imaging apertures by *A*, deflector plates by *D*, and mounting plates by *P*. The filament is designated by *F*, the cathode by *C*. This type of apparatus is most useful for a study of vibrational cross sections. (From Boness and Schulz, 1974.)

characteristic of boomerang states (see Sec. 2.1.2). The broad peaks in the vibrational cross section of N_2 in the energy ranges 7 to 8 eV and 13 to 14 eV are probably caused by higher excited states of N_2^-, with short lifetimes (see Sec. 2.1.1). The sharp peaks in the energy range 11 to 12 eV are caused by long-lived Feshbach resonances, as discussed in Sec. 2.1.3. Finally the regime between 15 and 30 eV exhibits an ultrabroad peak, resulting from doubly excited states, as discussed in Sec. 2.1.4.

2.1.6 Molecular Orbitals of Resonances. Table 1.1 shows the suggested molecular orbitals involved in the resonances discussed earlier, as well as the more conventional molecular orbitals of the states from which the resonances are derived ("parent states"). For completeness we have indicated in the table core-excited shape resonances associated with valence-excited states. These show a preferred decay into the parent valence-excited state, whereas their decay into the ground electronic state is highly unfavorable and they do not lead to a measurable vibrational cross section of the ground electronic state. They do, however, populate vibrational levels of the parent valence state (e.g., the $A^3\Sigma_u^+$ state of N_2).

On top of Table 1.1 we show the molecular orbitals in the proper sequence and the superscript on each orbital designation shows the number of electrons that can be accommodated. A checkmark in the table indicates that the particular orbital is fully occupied, a -1 indicates that one electron is missing from the particular orbital, thus creating a "hole." A 1 indicates that a single electron occupies the orbital. Table 1.1 is meant for orientation purposes, and only examples are given. For example, doubly excited states can be formed by taking electrons from other orbitals than those indicated. Thus there are many doubly excited states of similar energies, rather than just the one indicated. Similarly, one should be able to form a shape resonance by adding an electron to the $\sigma_u 2p$ orbital, rather than the $\pi_g 2p$ orbital, as indicated in Table 1.1. However occupation of the $\pi_g 2p$ orbital results in the *lowest* shape resonance.

2.2 Nonresonant Vibrational Excitation

Direct vibrational excitation can be induced in molecules by the dependence on internuclear separation of the interaction potential. The interaction potential outside the molecule itself (Takayanagi, 1967) can be written as

$$V = -\frac{qe}{r} - \frac{\mu e}{r^2} P_1(\hat{r} \cdot \hat{R}) - \frac{Qe}{r^3} P_2(\hat{r} \cdot \hat{R}) - \frac{\alpha e^2}{2r^4} - \frac{\alpha' e^2}{2r^4} P_2(\hat{r} \cdot \hat{R}) - \cdots \tag{1.1}$$

Table 1.1 Approximate molecular orbitals for N_2 and N_2^-

		Valence electrons							Rydberg electrons			
		KK	$(\sigma_g 2s)^2$	$(\sigma_u 2s)^2$	$(\sigma_g 2p)^2$	$(\pi_u 2p)^4$	$(\pi_g 2p)^4$	$(\sigma_u 2p)^2$	$(\sigma_g 3s)^2$	( )	State	Energy (eV)
N_2	Ground state	√	√	√	√	√	0	0	0	0	$X^1\Sigma_g^+$	0
N_2^-	Shape resonance	√	√	√	√	√	1	0	0	0	$^2\Pi_g$	1.7
N_2	Valence excited	√	√	√	√	−1	1	0	0	0	$A^3\Sigma_u^+$	6.2
N_2^-	Core-excited shape resonance	√	√	√	√	−1	1	1	0	0		8.2
N_2^+	Positive ion	√	√	√	−1	√	0	0	0	0	$X^2\Sigma_g^+$	15.6
N_2	Rydberg state	√	√	√	−1	√	0	0	1	0	$E^3\Sigma_g^+$	11.87
N_2^-	Feshbach resonance	√	√	√	−1	√	0	0	2	0	$^2\Sigma_g^+$	11.48
N_2^-	Core-excited shape resonance	√	√	√	−1	√	0	0	1	1		11.92
N_2	Doubly excited	√	√	√	−1	−1	1	1	0	0		~20
N_2^-	Doubly excited resonance	√	√	√	−1	−1	2	1	0	0		~20

Here qe is the net charge, r is the distance of the incident electron from the molecule, and R is the internuclear separation; μe is the electric dipole moment and the term containing μ is absent in homonuclear diatomic molecules. The third term, involving Qe, the quadrupole moment, characterizes the quadrupole interaction. All these three terms are the "electrostatic" terms in that they characterize the interaction of the incident electron and the unperturbed molecule. The next two terms, involving α, the spherically symmetric, and α', the nonspherical part, of the polarizability, are the "dynamic" terms. The latter involve the polarization of the molecule by the incident electron. In (1.1) $\hat{r}$ and $\hat{R}$ are unit vectors in directions r and R, respectively, and P_n are Legendre polynomials. The parameters α, α', μ, and Q are functions of R.

One of the problems encountered in the use of the r^{-4} potential is the need to cut off this potential since the r^{-4} law is applicable only at large distances from the molecule and does not hold near the molecule. Breig and Lin (1965) have addressed themselves to this problem, but in the end the analytic form of the cutoff remains somewhat arbitrary, as does the actual cutoff radius. This leads to great uncertainties in the calculated values for the vibrational cross sections via polarization interactions, which dominate in homonuclear diatomic molecules. Thus by the judicious choice of parameters, Breig and Lin (1965) and also Takayanagi (1965, 1967) were able to calculate vibrational cross sections that are in essential agreement with experiment in H_2 without invoking a resonance contribution, although the existence of a resonance is well established in this energy range. This will be discussed further in Sec. 4.1.

The interaction potential may be expanded as

$$V = V(R = R_e) + (R - R_e)\left.\frac{\partial V}{\partial R}\right|_{R=R_e} + \cdots \tag{1.2}$$

and higher-order terms are usually neglected. Combining this expansion with the harmonic oscillator assumption and the Born approximation leads to an approximate selection rule for vibrational transitions, namely,

$$\Delta v = 1$$

Thus we are led to the conclusion that direct excitation changes the vibrational quantum number by only unity, to a first approximation. Whenever experiment shows cross sections for higher vibrational states comparable in magnitude to those for the low ones, we are forced to abandon this simple approach. This occurs in many cases to be discussed later.

Itikawa (1974) has applied the Born approximation to a calculation of the vibrational excitation and he has been able to relate the vibrational cross sections by electron impact to quantities provided by optical spectroscopy (i.e., infrared absorption and Raman intensity). This is a useful development at this stage, especially for polyatomic molecules, since it provides an order-of-magnitude estimate for various cross sections. Close-coupling methods have also been used for the calculations of vibrational cross sections, and this study (Itikawa and Takayanagi, 1969) has provided a fruitful comparison with the Born approximation.

3 VIBRATIONAL EXCITATION AT LOW ENERGIES IN N_2, CO, CO_2 (BOOMERANG RESONANCES)

We now return to a more detailed discussion of vibrational excitation in several molecules, confined to the low-energy regime. For operation of gas lasers this is the most important regime, covering the range from 0 to about 4 eV. Vibrational excitation in this energy range strongly influences the distribution function of electrons and often a significant portion of the power input goes into vibrational excitation.

In the present section we combine the discussion of N_2, CO, and CO_2 because the lowest shape resonances in these molecules have similar characteristics: in these three molecules the resonance has an intermediate lifetime and the vibrational excitation can be understood in terms of the boomerang model (see Sec. 2.1.2). Of these molecules only CO has a permanent dipole moment, whereas the transition dipole moment must be taken into account for CO_2. This influences the calculations of the direct contribution to the vibrational excitation and the behavior of the cross section away from the resonance region.

3.1 N_2 and CO via the Lowest Shape Resonance

Vibrational excitation by electron impact on N_2 and CO in the energy range of the lowest shape resonance is now well understood. Being isoelectronic, N_2 and CO exhibit similar behavior with respect to electron impact in this regime. The bulk of the vibrational cross section at low energies for either N_2 or CO proceeds via a shape resonance. In N_2 this shape resonance has a designation $^2\Pi_g$ and is centered around 2.3 eV, whereas in CO the resonance has a designation $^2\Pi$ and is centered around 1.7 eV. The energy dependence of the cross section for the first 10 levels exhibits an oscillatory behavior (with peaks shifting to higher energies for higher vibrational levels) that can be understood in terms of the boomerang model of Birtwistle and Herzenberg (1971).

Figures 1.6 and 1.7 show the results for N_2 and CO as obtained in an electron spectrometer. The absolute values quoted for N_2 are based on the trapped-electron experiment of Spence, Mauer, and Schulz (1972), which yields a total vibrational cross section (sum of all vibrational levels) at 2.5 eV of $6.0 \times 10^{-16}\,\text{cm}^2$. This value is about 10% higher than the value quoted by Engelhardt, Phelps, and Risk (1964) ($5.3 \times 10^{-16}\,\text{cm}^2$), which is obtained from an analysis of transport coefficients in swarm experiments. Preliminary experiments by Wong suggest that the cross sections of Fig. 1.6 should be multiplied by a factor of 2. Further work on the determination of the absolute values is in progress. The absolute values quoted for CO are taken from Ehrhardt et al. (1968) as listed by Kieffer (1973).

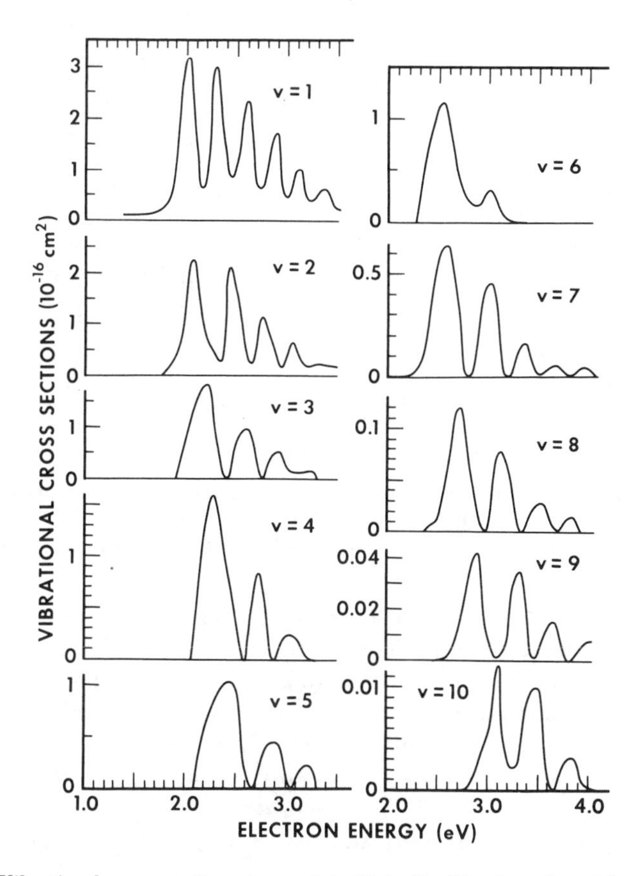

Fig. 1.6 Vibrational cross sections to $v = 1$ to 10 in N_2. The data plotted here are from Schulz (1964) for $v = 1$ to 6 and from Boness and Schulz (1973) for $v = 7$ to 10, normalized to the value given by Spence, Mauer and Schulz (1972). A digitized listing has been assembled by Kieffer (1973). The absolute values shown here may be too low by a factor of up to two (Wong, private communication).

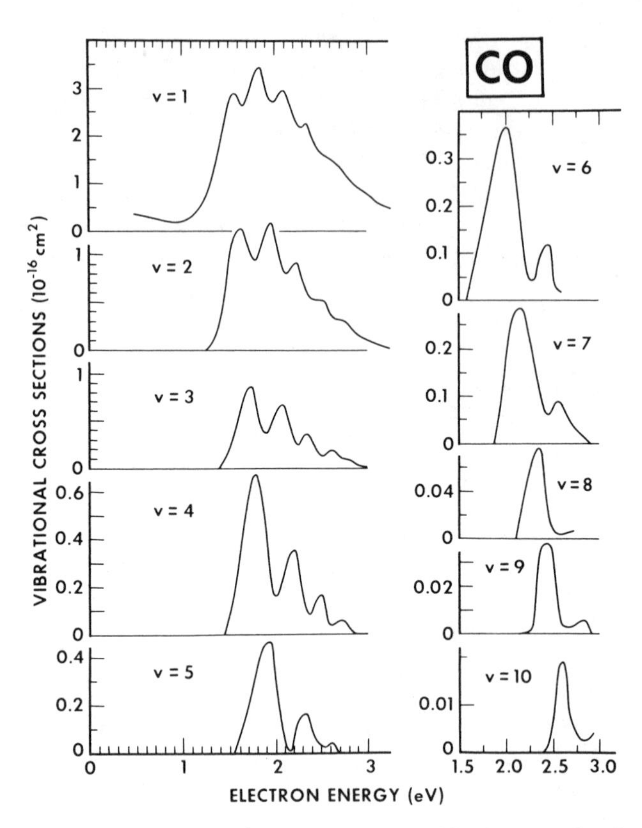

Fig. 1.7 Vibrational cross sections for $v = 1$ to 10 in CO. The data are taken from Ehrhardt et al. (1968) for $v = 1$ to 7 and from Boness and Schulz (1973) for $v = 8$ to 10. Absolute values are taken from Ehrhardt et al. (1968) as listed by Kieffer (1973).

3.1.1 Barrier Penetration in N_2 and CO. The relative cross sections for exciting different final vibrational states via a given resonance state are determined by the overlap of wavefunctions and by the probability of penetration through the centrifugal barrier. For high final vibrational states the matrix elements probably do not vary significantly in order of magnitude, and we can assume that the barrier penetration is the dominant factor (Boness and Schulz, 1973). The probability of penetration through a centrifugal barrier can be approximated by the following expression:

$$T \cong 4 \left(\frac{\epsilon}{E}\right)^{1/2} \alpha_l \qquad (1.3)$$

Here ϵ is the energy of the departing electron and E is the sum of the incident electron energy and the potential energy inside the molecule.

Equation 1.3 is valid only if $\epsilon \ll E$, which is certainly satisfied when high vibrational states are excited. The factor α_l depends on the partial wave l, which is dominant. Blatt and Weisskopf (1952) list the expressions for the factor α_l for different partial waves l. The partial wave involved in the case of N_2 is a d-wave ($l = 2$) and in the case of CO it is predominantly a p-wave ($l = 1$). With the restriction that $R(2m\epsilon/\hbar^2)^{1/2} < 1$, we can write the proportionalities

$$\begin{aligned} \alpha_1 &\sim \epsilon \qquad \text{for } l = 1 \\ \alpha_2 &\sim \epsilon^2 \qquad \text{for } l = 2 \end{aligned} \tag{1.4}$$

R is a cutoff radius. The transmission coefficients then satisfy the proportionality

$$T_1 \sim \frac{\epsilon^{3/2}}{E^{1/2}} \qquad \text{for CO} \tag{1.5}$$

$$T_2 \sim \frac{\epsilon^{5/2}}{E^{1/2}} \qquad \text{for } N_2 \tag{1.6}$$

The variation of $E^{1/2}$ is small over the energy range of interest, and the dominant variation for the transmission coefficients comes from the variation of the energy of the departing electron ϵ. We plot in Fig. 1.8 the magnitude of the vibrational cross section at the first peak as a function of ϵ, on a log–log scale. The energy of the departing electron ϵ is the difference between the energy of the first peak in the vibrational excitation and the energy of the vibrational level, computed from the spectroscopic constants given by Herzberg (1950).

Straight lines with a slope of $\frac{3}{2}$ (applicable to CO) and a slope of $\frac{5}{2}$ (applicable to N_2) are drawn through the experimental points. To the extent that the experimental points lie on a line with the proper slope, we find agreement between the theory presented here and the experiment. That the lowest point of the N_2 curve does not lie on the straight line with slope $\frac{5}{2}$ is not too significant, for the cross section for $v = 10$ is small and it is difficult to determine the exact energy for the first peak of the cross section in Fig. 1.6. This energy must be known in order to determine ϵ.

The ratio of cross sections to high vibrational states seems to be dominated by the quantum mechanical penetration through the centrifugal barrier, which the departing electron must overcome in order to leave the molecule. The extrapolation of plots similar to Fig. 1.8 can provide the magnitudes of the cross sections for higher-lying vibrational levels that are not accessible to measurement owing to a lack of sufficient detection sensitivity. Such an extrapolation can be used provided that the levels are energetically accessible via a resonance mechanism and provided that the

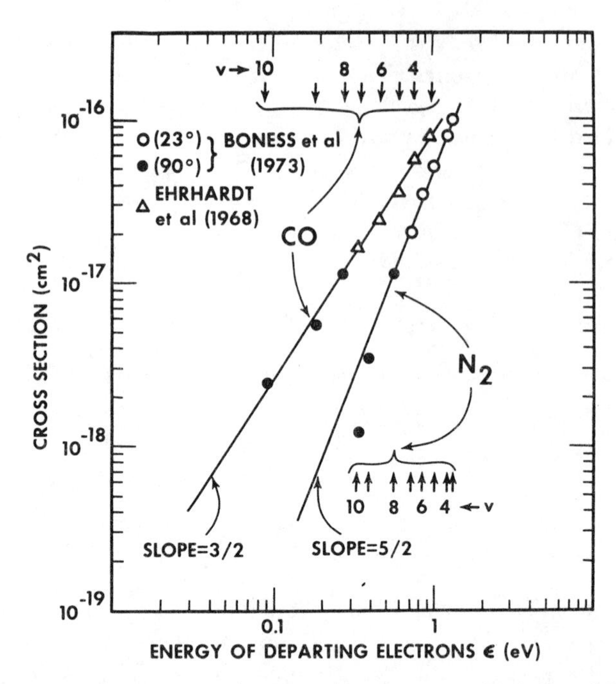

Fig. 1.8 Magnitude of the first peak in the vibrational cross section in N_2 and CO versus the energy of the electron after the collision ϵ. Under the assumptions discussed in the text, such a plot should give a straight line for high vibrational states, with a slope of $\frac{3}{2}$ for CO and $\frac{5}{2}$ for N_2. The straight lines have the slopes indicated in the figure. (From Boness and Schulz, 1973.)

approximations used in the derivation of the formulas are satisfied. The barrier penetration mechanism always leads to a higher peak cross section for the lower levels and thus no inversion of vibrational levels can take place in a gas discharge, in which the electron distribution extends over a significant energy range. The shift in the energies of the peaks of the cross sections shown in Fig. 1.6 could be used, in conjunction with electron beams, to obtain an inversion in vibrational populations. However this is a difficult proposal to implement.

3.2 N_2 and CO via Direct Processes

The threshold region for vibrational excitation in N_2 and CO is shown in Figs. 1.9 and 1.10. The most reliable experimental determinations for this regime are probably those deducing the cross sections from transport coefficients (Engelhardt, Phelps, and Risk, 1964, for N_2; Hake and Phelps, 1967, for CO). These have been discussed in summary form by Phelps

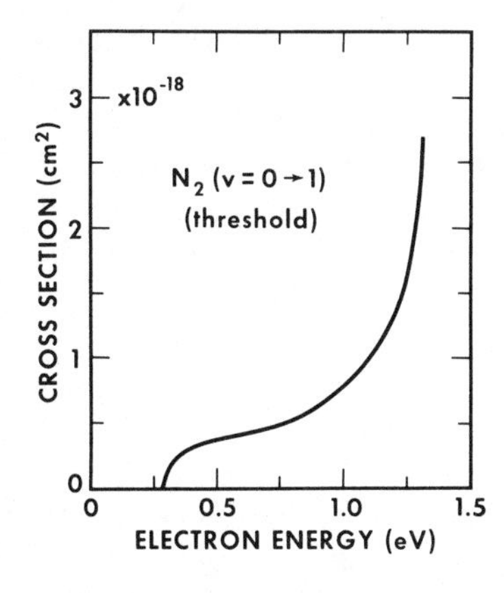

Fig. 1.9 Vibrational excitation in N_2 near threshold, as derived from an analysis of electron transport coefficients. (From Engelhardt, Phelps, and Risk 1964.)

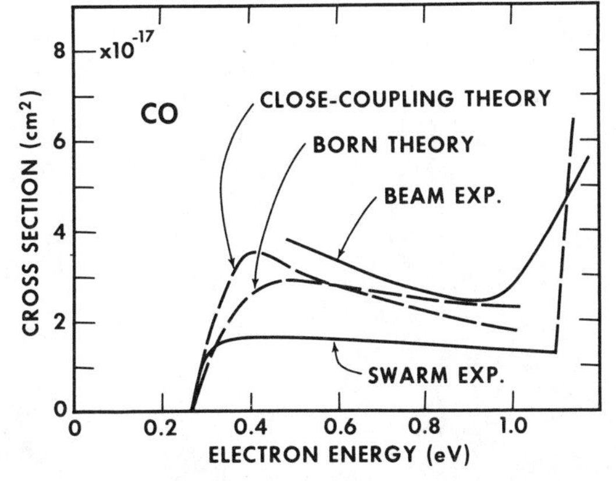

Fig. 1.10 Vibrational excitation cross sections for CO near threshold. The evaluation of the cross section from transport coefficients ("swarm exp") was done by Hake and Phelps (1967) and the cross section from the beam experiment is due to Ehrhardt et al. (1968). The calculations in the Born approximation and in close coupling have been done by Itikawa and Takayanagi (1969).

(1968), and it is pointed out that this method can give reliable values for cross sections near the vibrational threshold. However in both N_2 and CO the cross sections measured by electron beam experiments are higher (by a factor of 2 or more) compared to the cross sections deduced from transport coefficients. The reason for this discrepancy is not known.

Although we are discussing the threshold region for vibrational excitation in N_2 under the heading of "direct" processes, it is by no means certain that direct processes make a major contribution to vibrational excitation near threshold, and possibly the tail of the resonance may be dominating, as pointed out earlier. It should be noted that the absence of a permanent dipole in N_2 makes the cross sections exceptionally small (~0.3% of the peak value) near threshold.

For N_2 Phelps (1968) points out that the resonance theory of Chen (1966) actually agrees with the vibrational cross section near threshold, which is shown in Fig. 1.9. Phelps therefore suggests that no direct component of significant magnitude need be invoked for the interpretation of the vibrational cross section near threshold in N_2. A recent calculation of Chandra and Temkin (1976), however, does not support this contention. Their calculation of the resonant portion of the phase shift leads to a cross section about an order of magnitude smaller near threshold than the cross section of Engelhardt et al. (1964) shown in Fig. 1.9.

In the case of CO there is again a serious discrepancy between the experimental cross sections available to date. The analysis of transport coefficients (Hake and Phelps, 1967), which should be rather reliable in the threshold region, yields a cross section about one-half that obtained from beam experiments (Schulz, 1964; Ehrhardt et al., 1968), as shown in Fig. 1.10. Until this discrepancy is completely resolved it is difficult to test the theory in CO, since the discrepancy is of just the same order as the difference between the theory involving the dipole only and the theory

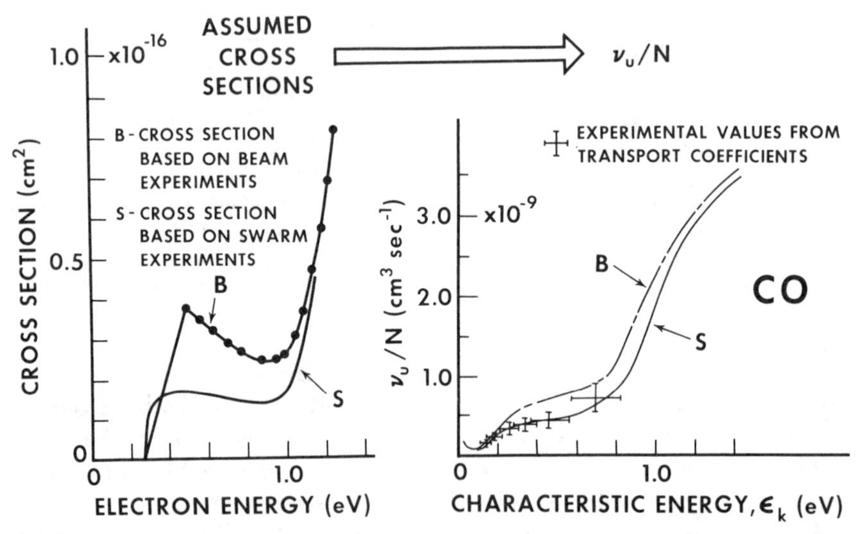

Fig. 1.11 Test of vibrational cross sections near threshold in CO. On the left side we show two curves: curve B is based on beam experiments and curve S is based on swarm experiments. On the right we show the coefficient ν_u/N, calculated for curve B and S on the left. Only curve S agrees with the experiment which is shown by points with error bars. (From Phelps unpublished.)

involving dipole plus polarization. The contribution of the tail of the resonance to the threshold region in CO has not yet been examined.

An attempt at resolving the discrepancy in experimental results has been made by Phelps (private communication) and Fig. 1.11 shows the argument. On the left side of Fig. 1.11, two "trial" cross sections are shown: (1) Curve *B* is based on electron beam data. The curve follows the cross section measured by Ehrhardt et al. (1968) down to 0.5 eV and then connects to the threshold of vibrational excitation with a straight line whose slope is somewhat larger than that measured by Burrow and Schulz (1969). (2) Curve *S* is based on the swarm data of Hake and Phelps (1967) at low energies and connects to the data of Ehrhardt et al. (1968) above about 1 eV. Using these two trial cross sections, Phelps derives the energy exchange collision frequency ν_u/N for each of these assumed cross sections, resulting in curves *B* and *S* on the right side of Fig. 1.11. Also shown on the right side of Fig. 1.11 are the experimental points for ν_u/N (including error bars) from experiments. Even within the conservative error bars shown in the figure, the swarm experiments clearly favor the trial cross sections marked *S*. The error bars conform to the empirical consistency of the experiments that lead to the calculation of ν_u/N. These consistency questions are discussed by Crompton et al. (1968).

We now summarize the results of various theories in the case of CO.

1. The best agreement with the experimental curve of Hake and Phelps (1967) (which must be preferred at this time because of the preceding argument) is obtained when polarization is neglected altogether, and the dipole contribution is calculated using the Born approximation. Phelps (1968) uses the Born formula of Takayanagi (1966), which relates the vibrational cross section to the infrared transition probability. The latter quantity is taken from Penner (1959). It is of course possible that the good agreement obtained in this manner is fortuitous. It should be noted that the use of the close-coupling theory for the calculating of the dipole contribution (Itikawa and Takayanagi 1969) underestimates the Hake and Phelps (1967) cross section by a factor of about 2.

2. Breig and Lin (1965) used the Born approximation together with the potential caused by the dipole and polarization. The cutoff parameters for the latter were varied within reasonable limits, leading to a range of total vibrational cross sections. Polarization accounts for between 25% and 50% of the total cross section. The results of these calculations agree with the beam experiments but are a factor of about 2 higher than the preferred cross sections derived from transport coefficients.

3. Itikawa and Takayanagi (1969) extended the considerations of Breig and Lin (1965) by testing whether the Born approximation can correctly

predict the vibrational cross section at low energies. By comparing the results of their close-coupling calculation with the results of the Born approximation with the same cutoff parameters, they find the following:

a. In the case of a pure dipole potential the Born approximation gives a vibrational cross section about twice that obtained for close coupling near threshold. At higher energies about 1.5 eV above threshold, the disagreement diminishes.

b. In the case of a pure polarization interaction the Born approximation gives only about 20% of the cross section obtained with close coupling, but both calculations depend strongly on the cutoff parameters used.

Figure 1.12 shows the components to the total vibrational excitation (dipole and polarization) as calculated by Itikawa and Takayanagi (1969) in the Born and close-coupling methods. The contribution of the tail of the $^2\Pi$ resonance was not included.

Until the experimental cross section is cleared up beyond doubt, it is difficult to evaluate the various theories. But clearly there is at present a degree of arbitrariness in the theories that makes it difficult to evaluate the contributions of the dipole, polarization, and resonance contributions to the threshold excitation.

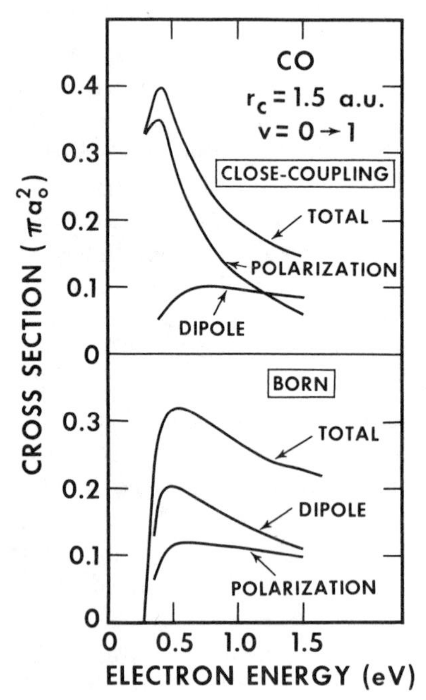

Fig. 1.12 Dipole and polarization components to the vibrational excitation for CO, in the Born and close-coupling theories. The resonance component to the vibrational excitation was not considered in these calculations. The cut off radius is taken to be 1.5 au. (From Itikawa and Takayanagi, 1969.)

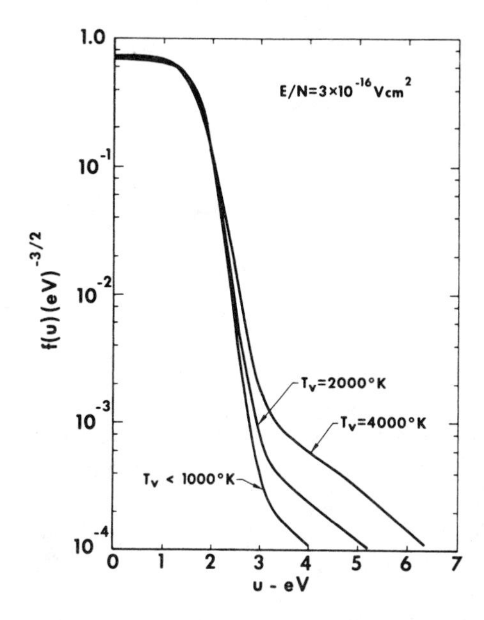

Fig. 1.13 Electron energy distribution functions in N_2 calculated for various vibrational temperatures. The energy distribution function $f(u)$ shows a precipitous drop at those energies at which the vibrational cross section becomes large. The electron energy is designated by u, and E/N is the ratio of the electric field to the gas density. (From Nighan, 1970.)

3.2.1 Effect of Vibrational Cross Sections on Electron Distribution Functions in Discharges. The large size of the vibrational cross sections in N_2 exerts a strong influence on the electron energy distribution function in nitrogen discharges by acting as a "barrier" for the gain in energy. Electrons thus experience difficulty in reaching energies in excess of about 2.5 eV so long as the vibrational temperature is low. This is shown in Fig. 1.13 for the curve marked $T_v < 1000°K$. When the vibrational temperature T_v is increased, superelastic collisions (in which electrons *gain* energy in collisions with vibrationally excited molecules) become important and the high-energy tail of the distribution function in Fig. 1.13 increases. The high-energy tail also increases when E/N (the ratio of the electric field to the gas density) is raised but the sharp cutoff is still present up to an E/N of 1.5×10^{-15} V cm^2, which corresponds to an average energy of 2.25 eV (Nighan, 1970). The value of E/N of 3×10^{-16} V cm^2, for which Fig. 1.13 is applicable, corresponds to an average electron energy of 0.72 eV.

3.3 CO_2 via the First Shape Resonance

Figure 1.14 shows a set of vibrational cross sections as derived by Lowke, Phelps, and Irwin (1973) from transport coefficients. It is evident from the energy dependence of the fundamental modes that two regions of the energy scale must be discussed separately. Below about 3 eV the excitation processes are direct: the optically allowed modes (001 and 010 modes)

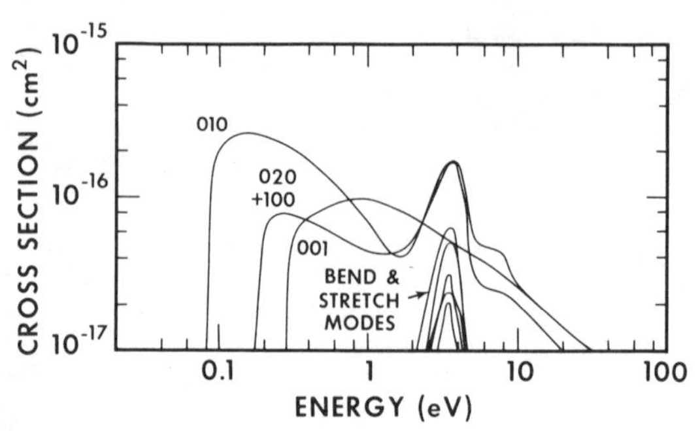

Fig. 1.14 Vibrational cross sections in CO_2 derived from electron transport coefficients. (From Lowke, Phelps and Irwin, 1973.)

become excited by the dipole interaction resulting from the transition dipole moments associated with these modes. This was first postulated by Andrick, Danner, and Ehrhardt (1969), who found that the electron scattering leading to the 001 and 010 excitation exhibits pronounced forward peaks. In fact, this dominant forward scattering persists up to about 20 eV. This finding has been put on a theoretical basis by Itikawa (1971). The excitation of the 100 optically forbidden mode below the resonance region exhibits a more isotropic angular distribution of electrons and it is attributed by Itikawa (1971) to a polarization interaction.

In the region of the first shape resonance, which is centered near 3.8 eV, the energy loss spectrum actually shows about 25 energy loss peaks, which are reproduced in Fig. 1.15. Boness and Schulz (1974) attribute these

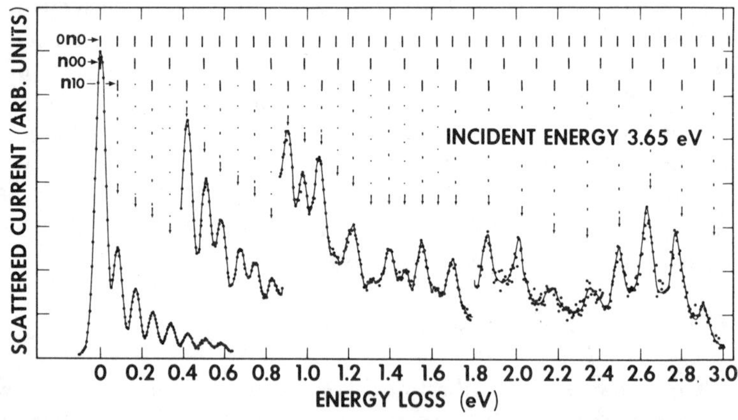

Fig. 1.15 Energy-loss spectrum in CO_2 at an angle of observation of 90° and an incident energy of 3.65 eV. The calculated positions for the vibrational progressions $0n0$, $n00$, and $n10$ are shown on top of the figure. The observed spectrum is interpreted in terms of two series, namely, $n00$ and $n10$. (From Boness and Schulz, 1974.)

energy loss peaks to two series, namely (n00) and (n10). The latter series is longer than the first.

The energy dependence of selected vibrational cross sections in the region of the resonance is shown in Fig. 1.16. Fine structure is evident for each

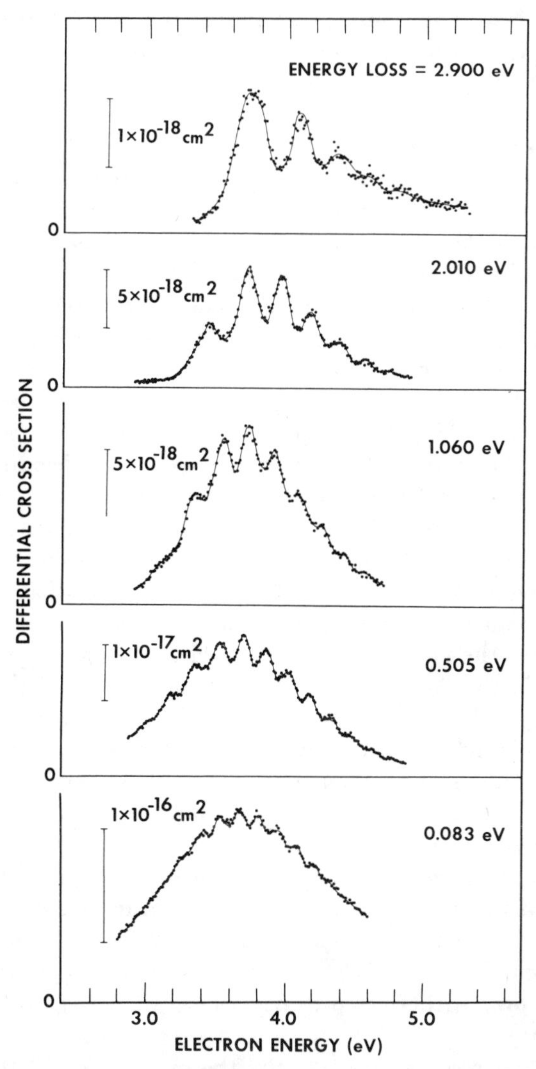

Fig. 1.16 Energy dependence of the differential cross sections for five different energy loss processes in CO_2. The approximate magnitudes of the cross sections are shown by the vertical bars. Note that the positions of the peaks shift on the energy scale. (From Boness and Schulz, 1974.)

mode shown. This fine structure shifts toward higher energies as the magnitude of the energy loss increases. These features are reminiscent of the vibrational excitation in N_2 and CO discussed in the previous section, and in particular one should recall the interpretation in terms of the boomerang model (Sec. 2.1.2).

Boness and Schulz (1974) have suggested that the interpretation of the resonance region in CO_2 around 3.8 eV should proceed in a manner analogous to that advanced for N_2 and CO.

3.3.1 Boomerang Model in Two Dimensions. Propagation of the nuclear wavepacket within the potential well of the N_2^- ion is adequately described in terms of a single coordinate, the internuclear separation. Application of the boomerang model to a triatomic molecule is complicated by the increased number of normal nuclear modes. Linear triatomic systems, in general, possess four normal nuclear modes, symmetric stretch, antisymmetric stretch, and two bending modes, the latter being degenerate. We consider the propagation of the nuclear wavepacket to be expressed as a function of only two coordinates, symmetric stretch and bending.

The variation of the potential energy for the lowest shape resonance in CO_2, $^2\Pi_u$, as a function of the symmetric stretch and bending coordinates has been calculated by Claydon et al. (1970) and their results are given in Fig. 1.17. The $^2\Pi_u$ state splits into two components ($^2A_1$ and $^2B_1$) as the CO_2^- ion departs from a linear configuration. In each portion of Fig. 1.17 the variation of potential energy is calculated assuming that the remaining coordinates retain their equilibrium configurations appropriate to CO_2.

Consider now the propagation of the nuclear wavepacket with respect to the two coordinates. At the instant of electron attachment the derivative of potential energy of the symmetric-stretch coordinate in the CO_2^- potential curve is negative and the nuclei immediately roll down along this coordinate. However for bending, the derivative is zero at 180° where attachment occurs and the nuclei are in a situation of unstable equilibrium. Because of the thermal or zero-point vibrational motion, however, the equilibrium is disturbed and the nuclei very slowly roll down the valley of the $^2A_1$ curve or up the sides of the valley of the $^2B_1$ curve shown in Fig. 1.17. In short, the negative ion attempts to reach its equilibrium configuration by bending very slowly but moving along the symmetric-stretch coordinate relatively rapidly. This proposition has previously been advanced by Taylor (private communication).

Figure 1.18 represents schematically the propagation of the envelope of the nuclear wavepacket as a function of the symmetric stretch and bending coordinates in the potential well of the negative ion. The diagram has been constructed as an extension of the boomerang model, by assuming that only

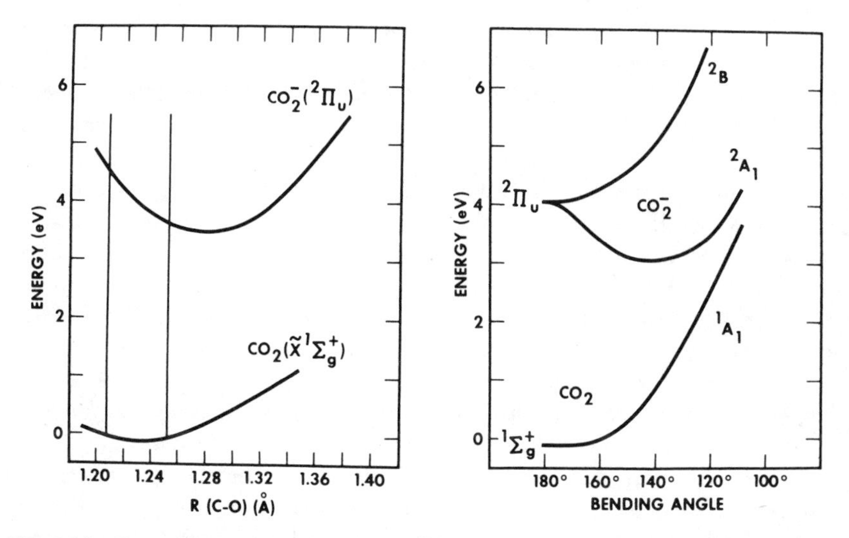

Fig. 1.17 Potential-energy variation for the $^2\Pi_u$ resonance of CO_2 versus symmetric stretch and bending coordinates. The bond angle is fixed at 180° for the left-hand plot; on the right-hand plot, the bond length is maintained at its equilibrium distance of 1.23 Å. (From Claydon, Segal, and Taylor, 1970.)

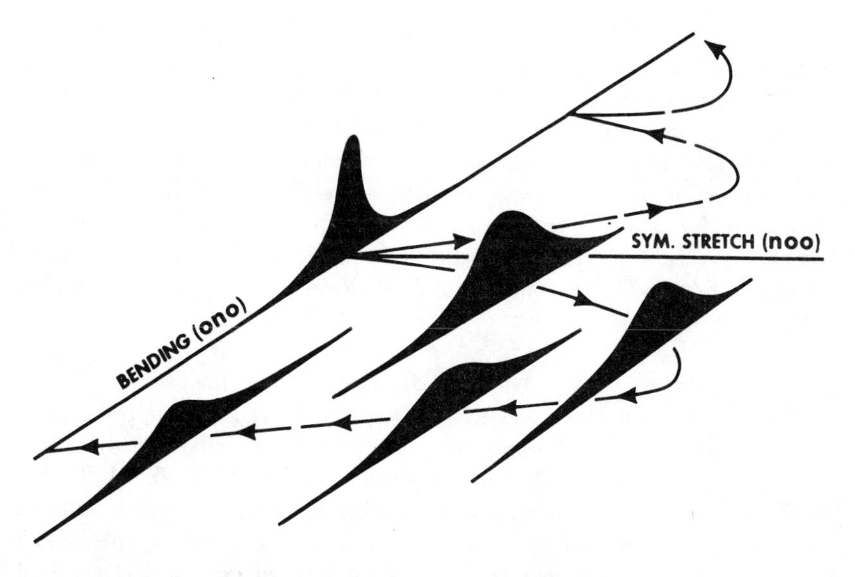

Fig. 1.18 Schematic diagram for the boomerang model in two dimensions, applicable to the case of CO_2. The nuclear wavefunction performs a single full excursion in the symmetric-stretch direction and a small excursion in the bending direction. (From Boness and Schulz, 1974.)

one reflection of the wavepacket occurs along the symmetric-stretch direction, after which the wavepacket is annihilated by autodetachment. Accompanying the excursion in the symmetric-stretch direction is a small excursion in the bending direction. The smallness of the bending excursion is a result of the anharmonic nature of the variation of the potential energy in the bending direction. Since the bending excursion is much less than one complete cycle, no reflection of the wavepacket occurs along this coordinate and all the oscillatory structure occurring in the energy dependence of the cross section (Fig. 1.16) is a consequence of the standing wave along the symmetric-stretch coordinate. Upper and lower dashed trajectories are drawn corresponding to the two possible directions for deformation of the bond angle from 180°. Two loops have been included in the upper trajectory. The second loop corresponds to two additional reflections of the nuclear wavepacket and has been included to indicate the development of the trajectory if additional oscillations occur along the symmetric-stretch coordinate. Commencing at the intersection of the axes and proceeding in the direction of the arrows, the time evolution of the envelope of the wave packet is indicated at a number of points along the lower trajectory. At the expense of completeness but in the interest of clarity, only sections through the wavepacket parallel to the bending coordinate are shown.

3.3.2 Decay of the Two-Dimensional Boomerang. The preceding discussion implies that the Franck-Condon overlap between the negative-ion state and the neutral ground state along the symmetric-stretch coordinate encompasses many highly excited symmetric-stretch vibrational levels of the neutral ground state. However since only a small deformation of the bond angle occurs during the lifetime of the resonance, the bending coordinate overlap includes very few bending vibrational levels. Thus copious symmetric-stretch excitation should occur accompanied by little or no bending excitation. These conclusions are in keeping with our interpretation of the energy loss spectrum.

3.3.3 Relationship to Angular Distributions. Inspection of the angular-distribution measurements of Danner (1970) lends support to the model advanced earlier. Danner measured angular distributions for those electrons having lost energies equal to 0.083, 0.167, 0.255, 0.339, 0.422, 0.505, and 0.591 eV at impact energies within the domain of the resonance. These loss processes correspond to the first seven peaks of the energy loss spectrum of Fig. 1.15. Their measurements revealed two distinct types of angular distributions, and alternate energy loss processes exhibited identical distributions. The angular distribution of electrons having excited the 010 bending mode possessed an anomalous feature not exhibited by the other members of the same series, a sharp peak in the forward direction. This peak arises from the direct interaction that couples via the transition dipole moment and that is unimportant for the higher-loss processes.

The angular distributions observed by Danner (1970) are sketched in Fig. 1.19, which also gives the absolute values of the cross section for various excitation modes, as provided by different experiments. The direct component of the 010 loss process is indicated by the dashed portion of the $n10$ distribution.

Andrick and Read (1971) derived expressions for the angular distribu-

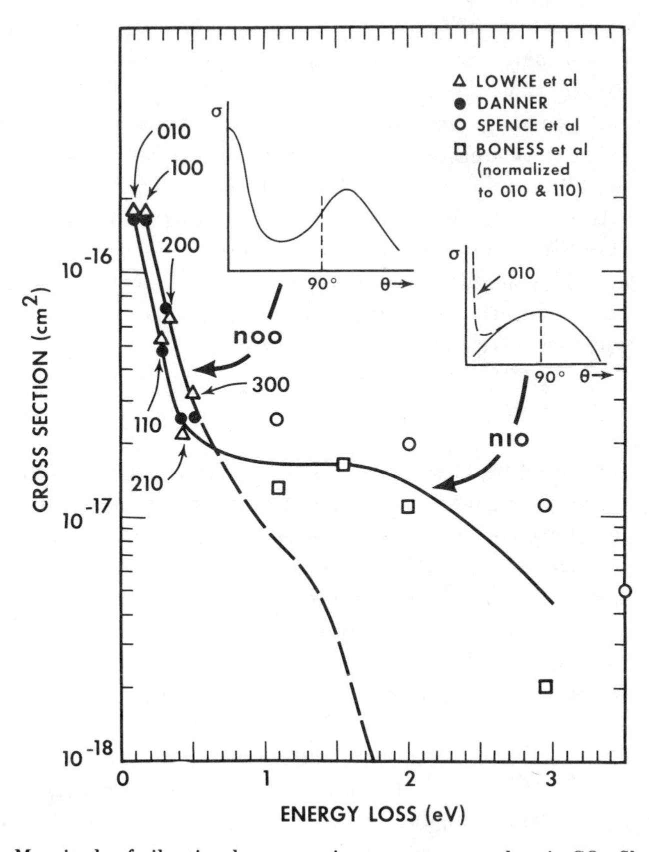

Fig. 1.19 Magnitude of vibrational cross sections versus energy loss in CO_2. Shown are the data of Boness et al. (1974), Lowke et al. (1973), Danner (1970), and Spence et al. (1972). The heavy lines connect points belonging to two series, $n10$ and $n00$. The dashed portion of the curve for the $n00$ series (0.6 to 1.7 eV) is based on the data of Boness et al. (1974), but this portion is afflicted with a large error. The angular distributions characteristic of each series were measured by Danner (1970) and they are sketched in. The angular distribution for the 010 mode is unique and is shown dashed. (From Boness and Schulz, 1974.)

tions of electrons resonantly scattered by molecules and demonstrated how these distributions depend on the vibronic symmetries of the molecular states. By adjusting the amplitudes and phases of the partial waves contributing to the particular excitation process, they were able to obtain excellent least-squares fits to the experimental data obtained for the 0.083 and 0.172 eV loss processes. The partial-wave composition was deduced from symmetry considerations assuming that the loss processes corresponded to excitation of the 010 and 100 modes, respectively. Energetically the excitation of the 100 mode could not be resolved from the 020 mode; however the excellence of the fit to the experimental angular distributions resolves this ambiguity and confirms the identification of the loss process as 100, which is in agreement with the interpretation given here.

The postulate that the two series $n10$ and $n00$ are excited is also consistent with the angular-distribution measurements. The series possess different vibronic symmetries that are constant within a particular series. The theoretically predicted distributions conform to the experimental measurements. Each member of a bending series, such as $0n0$, would possess different vibronic symmetry and therefore produce different angular distributions. The angular-distribution measurements thus offer strong support to the contention that higher bending modes are not excited since only two types of distribution are observed.

3.4 CO_2 via Direct Processes

As already pointed out in Sec. 3.3 there are inelastic processes for vibrational excitation below about 3 eV, which we attribute to direct excitation. Near the very threshold of excitation (i.e., somewhere between 5 to 40 meV above threshold), all fundamental vibrational modes can be excited (Stamatovic and Schulz, 1969). The ratio of cross sections near threshold is 2.3 : 0.9 : 1 for the modes 010 : 100 : 001. A transition dipole moment exists only for the 010 and 001 modes, and the Born approximation using the infrared transition probability predicts the ratio 010 : 001 fairly well. Thus one can conclude that the dipole interaction is responsible for the excitation of the 010 and 001 modes. The angular-distribution experiment of Andrick et al. (1969), who find a strong forward peak for these two modes (albeit at a considerably higher electron energy, i.e., 1.9 eV) confirms this viewpoint.

Since no transition dipole is involved in connection with the excitation of the 100, symmetric-stretch mode, other mechanisms must be invoked to explain its excitation near threshold. Itikawa (1970) invokes the polarization interaction for this purpose. This of course is subject to all the uncertainties involved in the use of the polarization interaction (see Sec. 3.2). Nevertheless the angular distribution for this mode, as measured by Andrick et al. (1969), is essentially isotropic, as is characteristic of the polarization interaction (see Itikawa, 1970). Also it has been shown for CO that the polarization interaction (in the close-coupling theory) leads to a much stronger decrease in cross section with electron energy compared to the dipole interaction. Thus we could explain why the 100 mode (being excited via the polarization interaction) is much weaker at 1.9 eV (Andrick et al., 1969) than the 010 and 001 modes, whereas the 100 mode is of comparable magnitude to the 010 and 001 modes near the threshold (Stamatovic et al., 1969).

The preceding model appears to explain the present data satisfactorily. For completeness, however, we should point out a suggestion advanced by Tice and Kivelson (1967), which is still untested in detail but may merit more serious consideration in the future.

Tice and Kivelson argue as follows: Although CO_2 has no permanent dipole moment, it does have instantaneous dipole moments arising from vibrational distortions. In particular, the low-frequency bending mode with an angular frequency $\omega = 12 \times 10^{13}\ \text{sec}^{-1}$ may give rise to an appreciable dipole moment. In the ground state the mean-square deviation of the O=C=O angle from 180° is θ^2, where

$$\theta^2 = \left(\frac{\hbar\omega}{2}\right)\left(\frac{k_\theta}{L^2}\right)^{-1}\left(\frac{1}{L^2}\right) \tag{1.7}$$

The C=O bond length is $L = 1.16$ Å and the force constant $(k_\theta/L^2) = 0.57 \times 10^5$ dyn/cm. Thus $\theta^2 = 72 \times 10^{-4}$. From infrared intensity measurements a dipole moment of 1.33 D can be assigned to the C=O bond in carbon dioxide. Tice and Kivelson arrive at a value for the root-mean-square dipole moment of approximately 0.16 D. Tice and Kivelson point out that electrons of high velocity ($v > 10^7$ cm/sec) pass through the region of the oscillating dipole so rapidly that they see a fixed dipole moment. The dipole moment being felt by a given electron depends on the phase of the molecular vibration at the time of encounter. On the other hand, electrons with velocities less than 10^6 cm/sec see the dipole moment averaged over many oscillations; the average dipole moment is zero. Electrons with velocities in the range from 10^6 and 10^7 cm/sec are subjected to a very complex effective interaction with the molecule.

Although Tice and Kivelson did not consider vibrational excitation by the process outlined here, it would appear that the argument can be readily extended to the excitation of vibrational modes that normally do not have a transition dipole moment (i.e., those modes that are not infrared-active).

4 VIBRATIONAL EXCITATION AT LOW ENERGIES IN H_2, N_2O, AND H_2O (SHORT-LIVED RESONANCES)

In this section we discuss those molecules whose low-lying shape resonance is of such a short lifetime that very little motion of the nuclei can take place within the lifetime of the resonance. The short lifetime of the resonance leads us immediately to certain conclusions regarding the properties of the vibrational cross sections; there can be no reflected component of the nuclear wavefunction set up in the negative-ion system because the wavepacket formed in this system does not propagate appreciably before autodetachment. This leads to the absence of fine structure in the vibrational cross sections and a relatively broad (structureless) curve. The short lifetime prevents significant internuclear motion during the lifetime of the resonance and thus decay to the lowest vibrational states is strongly favored. Higher vibrational states are expected to be only weakly excited. The molecules that fit into this category are H_2, N_2O, and H_2O.

4.1 H_2

Vibrational excitation via the lowest-lying shape resonance $^2\Sigma_u^+$ of H_2 has been discussed in great detail in the recent literature (e.g., Schulz, 1973). Its relationship to dissociative attachment, elastic scattering, and rotational excitation has provided a fertile field for the application of the impulse theory. Figure 1.20, given here for completeness, shows the expected behavior; no fine structure is present in the vibrational excitation and the ratio of excitation of higher modes (e.g., $v = 2$) to that of lower modes (e.g.,

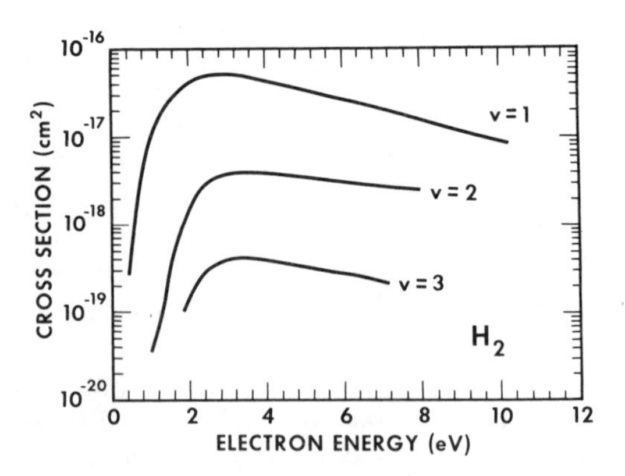

Fig. 1.20 Energy dependence of vibrational cross sections for $v = 1$ to 3 in H_2. (From Ehrhardt et al. (1968), as listed by Kieffer, 1973.)

$v = 1$) is small (~0.1). The formation of H^- via the $^2\Sigma_u^+$ resonance is also small (~10^{-21} cm^2) because the survival probability is very small.

Figure 1.21*a* shows the energy loss spectrum for H_2 near the center of the resonance range. Pure rotational excitation as well as rotational-vibrational excitations are clearly resolved.

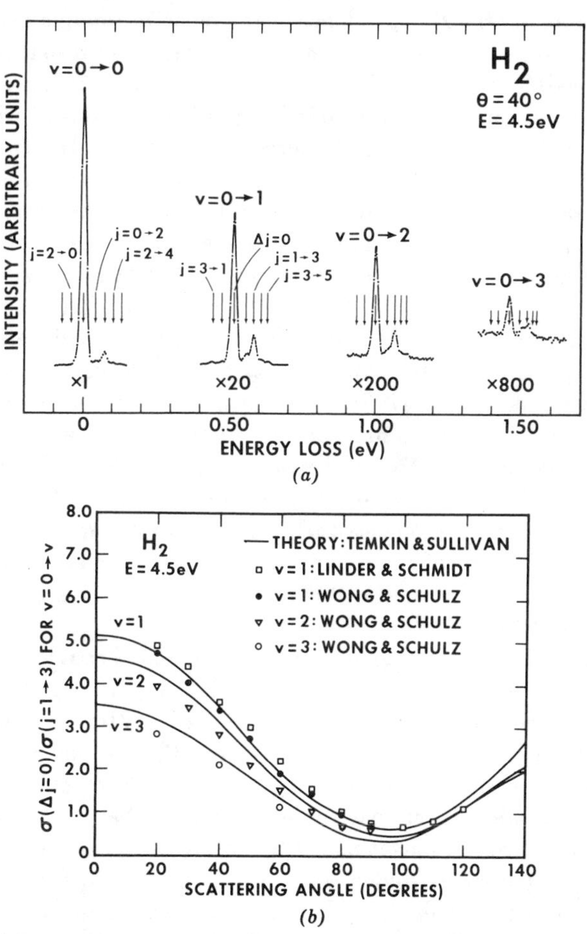

Fig. 1.21 (*a*) Energy-loss spectrum for H_2 at an incident energy of 4.5 eV and an angle of observation of 40°. Vibrational excitation to $v = 1$ to 3 is shown in three data blocks and the elastic peak is marked $v = 0 \rightarrow 0$. The vertical bars in each data block indicate the energy levels for allowed rotational transitions. The selection rule is $\Delta j = 0$ and $\Delta j = \pm 2$. (From Wong and Schulz, 1974.) (*b*) Angular dependence of the cross-section ratio $\sigma(\Delta j = 0)/\sigma(j = 1 \rightarrow 3)$ for the vibrational levels $v = 1, 2, 3$ in H_2, obtained from data such as shown in Fig. 2.21*a*. Shown are the experimental results of Wong and Schulz (1974) for $v = 1$ to 3 and of Linder and Schmidt (1971b) for $v = 1$. The solid lines represent the calculations of Temkin and Sullivan (1974) for $v = 1$ to 3.

The direct component and the resonance components of vibrational excitation are not separated in energy and thus we cannot attempt to discuss them separately: They are intermixed from an experimental viewpoint and reliance has to be placed on theory. It should be noted that Takayanagi (1967 and references therein) was able to explain the magnitude of the vibrational cross section using the R-dependence of the polarization in a distorted-wave calculation, but this alone does not assure that a nonresonant process is solely responsible for the excitation mechanism. Rather it appears very well established from the recent work of Chang (1974), combined with previous work, especially of Abrams and Herzenberg (1969), Henry and Chang (1972), and Temkin and Sullivan (1974), that a combination of direct and resonant processes must be invoked for interpreting the cross section for $v=1$ (and possibly $v=2$) and that the excitation of $v=3$ is purely resonant. This can best be seen from Fig. 1.21*b*, which shows the ratio $\sigma(\Delta j=0):\sigma(j=1\rightarrow 3)$ as a function of angle of scattering. This ratio shows a different angular behavior for different final vibrational states. Shown in Fig. 1.21*b* are the experimental data of Wong and Schulz (1974) and of Linder and Schmidt (1971a) in comparison with the theory of Temkin and Sullivan (1974). As pointed out by Chang (1974) whose calculations bear a great similarity to those of Temkin et al. (1974), the theoretical curve for $v=3$ is characteristic of a pure resonance, without any direct component. The pure resonance theory is due to Abrams and Herzenberg (1969). The agreement of the curve for $v=3$ with the resonance theory of Abrams et al. shows that no direct component of scattering need be invoked for the vibrational excitation to $v=3$.

Two effects cause a departure in the angular behavior of the ratio $\sigma(\Delta j=0)/\sigma(j=1\rightarrow 3)$ as one goes from $v=3$ to $v=1$. Most important is the contribution of the direct component of scattering (Chang, 1974). Temkin and Sullivan (1974) point out that, in addition, one needs to include the dependence of the vibrational wavefunction on the rotational quantum number. One would expect that the excitation of the $v=2$ level could be described by a purely resonant model. However this does not seem to be the case since the angular dependence for $v=2$ (plotted in Fig. 1.19) shows a significant deviation from the purely resonant scattering of $v=3$.

The resonant contribution to the excitation of vibration was originally proposed by Bardsley, Herzenberg, and Mandl (1966) via the $(1s\sigma_g)^2(2p\sigma_u)^2\Sigma_u^+$ state of H_2^-. Such an assignment also comes naturally from the results of the frame transformation theory (Henry and Chang, 1972) by showing a large derivative in the p_σ phase shift.

Despite the relative simplicity of the H_2 molecule it has been difficult to pin down the physical processes leading up to vibrational excitation in H_2, because direct and resonance contributions are not clearly separated in energy. Nevertheless the need for invoking a low-lying shape resonance is compelling, especially when additional channels of decay are considered, namely, rotational excitation and dissociative attachment, via the $^2\Sigma_u^-$ resonance. The latter process and its isotope effect have been recently reviewed (Schulz, 1973).

4.2 N_2O

We present here evidence that short-lived resonances are present in some triatomic systems as well, and the example that can be given is the *second* shape resonance in N_2O. Unfortunately the first shape resonance, which is located below 0.5 eV (Chantry, 1969; Zecca et al., 1974), is placed at such low energy that a study, using electron spectrometers, is difficult and has not yet been undertaken. Thus we must be content with a study of the $^2\Sigma^+$ resonance, located near 2.3 eV.

Figure 1.22 shows the energy loss spectrum taken at the center of the $^2\Sigma^+$

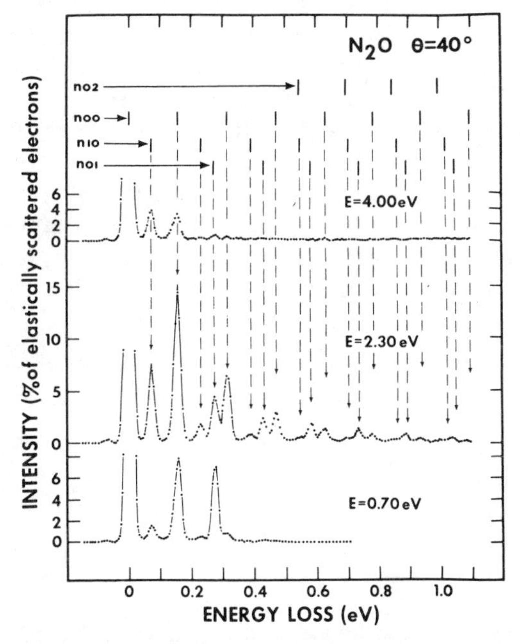

Fig. 1.22 Energy loss spectra obtained at an angle of observation of 40° and incident energies of 0.70, 2.30 and 4.00 eV, respectively, in N_2O. On the top of the figure the calculated positions for the vibrational progressions $n00$, $n10$, $(n-3)02$, $n01$ are indicated. The small peaks to the left of the zero-energy peak result from the deexcitation of the N_2O molecules which are in the 010 vibrational state (superelastic collisions). (From Azria, Wong, and Schulz, 1975.)

resonance and also off resonance (at 0.7 and 4.0 eV). At an incident energy of 2.30 eV the vibrational cross sections (except for the fundamental modes) are almost an order of magnitude larger than at the other two energies shown in Fig. 1.22. In addition, the energy loss spectrum at 2.30 eV shows many more vibrational modes than the spectra at 4.0 and 0.70 eV. These observations show the contribution of the $^2\Sigma^+$ resonance to the vibrational excitation. It should be noted that the excitation to the *fundamental* modes shows a significant nonresonant contribution at low energies, which is especially pronounced for the 001 mode.

The data of Fig. 1.22 also suggest that at least three separate series of vibrational modes are populated via the resonance. The location of the energy loss peaks is in good agreement with the location of the vibrational series, $n00$ (symmetric-stretch mode), $n10$ (symmetric-stretch mode plus 1 quantum of bending), and $n01$ (symmetric-stretch mode plus 1 quantum of asymmetric-stretch mode). The positions of these levels, calculated from the fundamental frequencies and the anharmonic coefficients given by Herzberg (1950), are indicated at the top of Fig. 1.22. However for energy losses equal to or larger than the energy of the 310 mode there are energy coincidences with the $n02$ series (symmetric-stretch mode plus 2 quanta of asymmetric stretch mode). Thus one cannot resolve experimentally energy loss peaks resulting from the excitation of the $n10$ mode and those of the $(n-3)02$ mode, for $n > 3$.

Figure 1.23 shows the energy dependence of the differential vibrational cross sections at 40° for the excitation of the 110 and 200 modes of N_2O. These curves, with a peak near 2.3 eV, exhibit no fine structure, as would be expected for short-lived resonances.

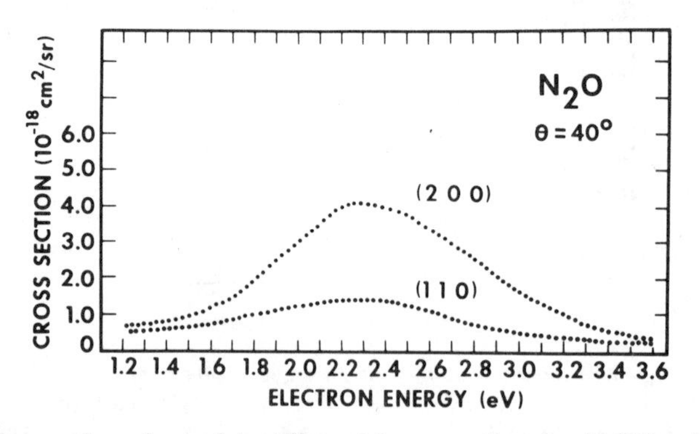

Fig. 1.23 Energy dependence of the differential cross sections for the 200 and 110 modes in N_2O. The absence of any structure in these curves points to the possibility that a short-lived resonance is involved in populating the vibrational states in this energy range. The estimated error for the cross section scale is about 35%. (From Azria, Wong, and Schulz, 1975.)

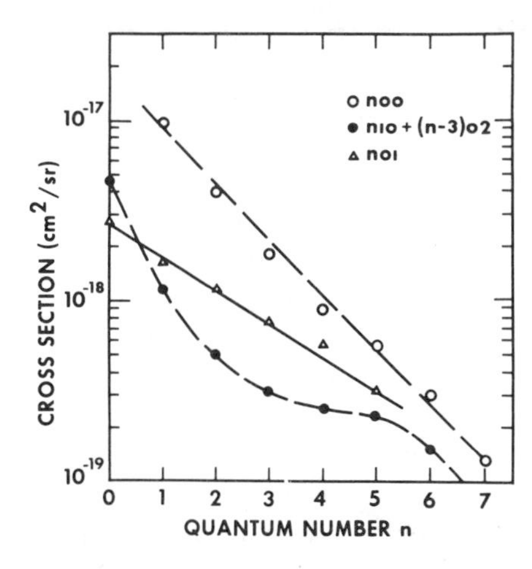

Fig. **1.24** Magnitude of vibrational cross sections for the series $n00$, $n01$, and $[n10 + (n-3)02]$ versus the quantum number n at an energy of 2.3 eV and a scattering angle of 40° in N_2O. The two latter series, $n10$ and $(n-3)02$, are not distinguishable for $n > 3$. The theory of Dubé and Herzenberg (1975) shows that both these modes contribute to vibrational excitation. (From Azria et al., 1975.)

Another prominent feature of the vibrational excitation processes in N_2O is the attenuation in amplitude of peaks that are members of the same series. Figure 1.24 shows the magnitude of the vibrational cross sections at 2.3 eV plotted against the quantum number n, for the three series $n00$, $n10$, and $n01$. For the $n00$ and $n01$ series the vibrational cross sections are reduced exponentially as n increases. This is not the case, however, for the $n10$ series, since this series has an admixture of the $(n-3)02$ series. Calculations have been performed by Dubé and Herzenberg (1975) within the framework of the "impulse approximation." By obtaining a good fit to the energy dependence for the vibrational modes 200 and 110 of Fig. 1.23, they establish that the lifetime of the N_2O^- $(^2\Sigma^+)$ resonance is indeed short. Their value for the width Γ is about 0.7 eV in the center of the Franck-Condon region and it decreases for larger internuclear separations. When the width is used to obtain the branching ratios of the vibrational series of Fig. 1.24, very good agreement results for the experimentally resolved series $n00$ and $n01$, and also for the sum of the $n10 + (n-3)02$ series.

4.3 H_2O

The energy loss spectrum for H_2O at 6 eV (Fig. 1.25) shows peaks caused by excitation of the (010) and (100 + 001) modes and a small contribution from a higher mode. The large size of the (100 + 001) cross sections (which cannot be resolved in this experiment) facilitates inversion in H_2O lasers since these are the upper laser states (Benedict et al., 1969). The energy dependence of the (010) and (100 + 001) modes (Fig. 1.26) shows a broad peak about 5 eV wide centered near 7.5 eV. This peak is attributed to one of two resonances (Seng and Linder, 1974) that escaped detection in transmis-

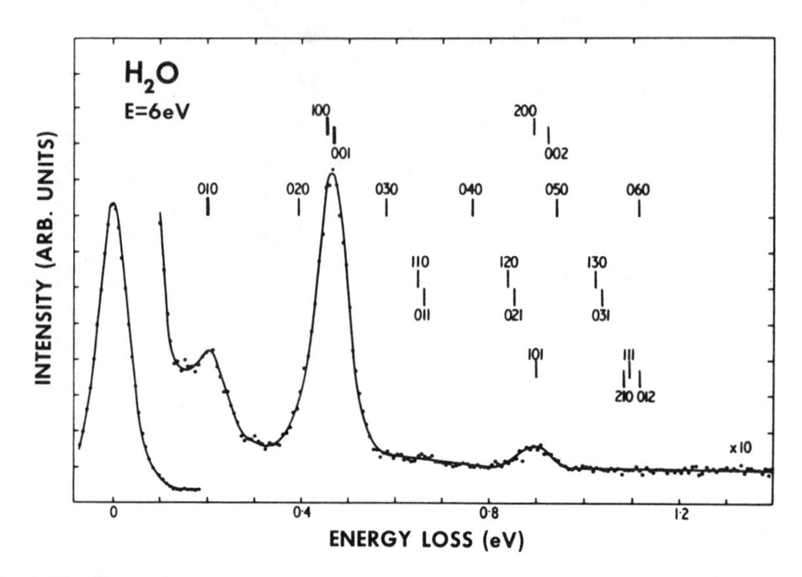

Fig. 1.25 Energy loss spectrum in H_2O at a collision energy of 6 eV and a scattering angle of 110°. The positions of possible vibrational modes are indicated by vertical bars. (From Seng and Linder, 1974.)

sion experiments of Sanche and Schulz (1973) because of their large width, and it is reminiscent of the case of H_2 and N_2O. Because of the short lifetime of the resonance very little internuclear motion can take place during the lifetime of the resonance and thus only the fundamental modes become appreciably excited. This is analogous to the dominance of the $v = 1$ excitation via the broad resonance in H_2.

At lower energies, below about 2 eV, Seng and Linder's (1974) cross section exhibits a rise, as shown in Fig. 1.26. This regime is attributed by Seng et al. to direct vibrational excitation, but it is expected that resonance processes of some sort will have to be invoked in order to explain all the experimental facts. The cross sections at these low energies have been calculated by Itikawa (1974) using the Born approximation, and the theoretical results are shown in Figs. 1.27 and 1.28. For the 010 mode Itikawa obtains a peak value about 4×10^{-17} cm^2. For the (100 + 001) modes the Born approximation gives a value about four times smaller, whereas the differential cross section of Seng and Linder (1974), taken at 90°, gives a value 50% larger than the 010 contribution. When the differential cross sections are integrated over angle, as has been done by Wong and Schulz (1975), it is found that the (010) cross section is of approximately the same magnitude as the (100 + 001) cross section (i.e., 6.3×10^{-17} cm^2 near 1 eV). Thus a discrepancy exists between theory and experiment in the low-energy

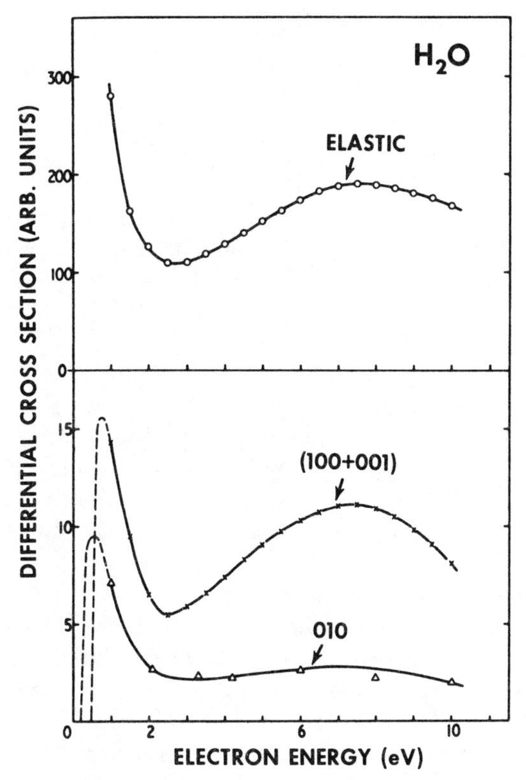

Fig. 1.26 Energy dependence of elastic and vibrational cross sections in H_2O. (From Seng and Linder, 1974.)

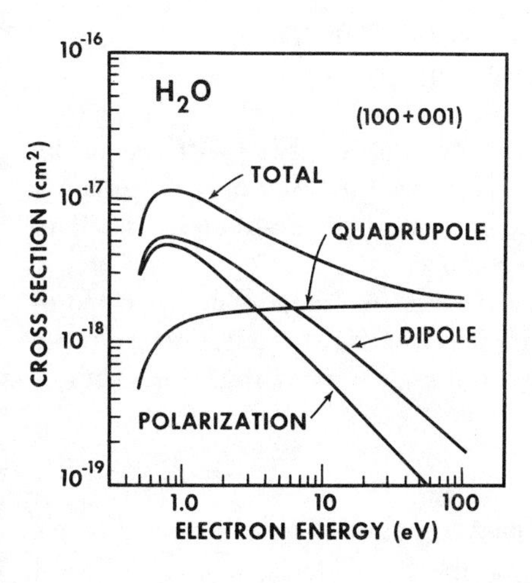

Fig. 1.27 Theory using the Born approximation for vibrational excitation in H_2O. Shown are the individual components as well as the total cross sections for the (100 + 001) modes. (From Itikawa, 1974.)

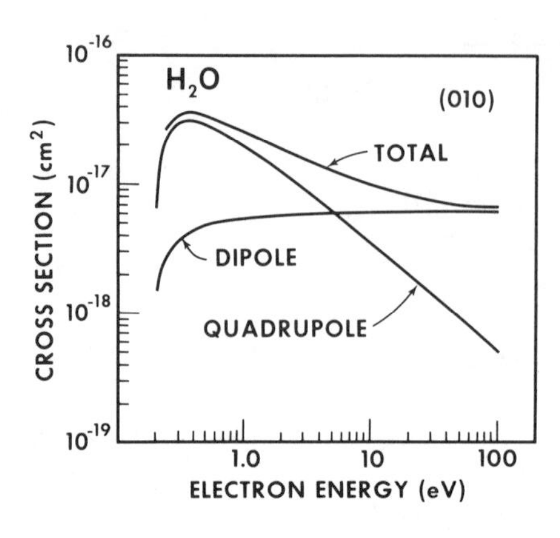

Fig. 1.28 Theory using the Born approximation for the (010) excitation in H_2O. Only the dipole and quadrupole components are considered. The total cross section is the sum of these. (From Itikawa, 1974.)

region. The angular distribution of electrons having excited the (100 + 001) modes also shows a discrepancy between experiment and theory: Whereas the calculation of Itikawa (1974) predicts a strongly forward-peaked distribution, the experiment of Wong et al. (1975) shows an isotropic distribution.

It has been pointed out by Wong and Schulz (1974b) that for those states of molecules that have a large dipole moment (e.g., the $^3\Pi$ state of CO) one observes resonances associated with this dipole. They labeled such resonances *dipole-dominated* resonances. It is entirely plausible that the low-energy region of H_2O (and also of HCl, HF, and so on) is affected by such a dipole-dominated resonance. If this turns out to be the case, then the Born approximation would not be valid and a resonance approach would have to be used.

5 VIBRATIONAL EXCITATION AT LOW ENERGIES IN O_2 AND NO (LONG-LIVED RESONANCES)

Some molecules such as O_2 and NO exhibit a resonance of long lifetime as the first shape resonance. There are other candidates for this category, possibly benzene, but this molecule has not yet been studied in sufficient detail to qualify for definite inclusion in this category.

As has already been pointed out, molecules in this category are expected to exhibit narrow, isolated spikes in their vibrational excitation and the energies of these spikes remain constant for all final vibrational states.

5.1 O_2

The lowest compound state of O_2 is the $X^2\Pi_g$ state. The vibrational levels $v'=0$ to $v'=3$ of $O_2^-(X^2\Pi_g)$ lie below the $v=0$ level of $O_2(X^3\Sigma_g^-)$ and cannot autodetach. These vibrational levels of O_2^- are therefore stable. For higher quantum numbers autodetachment can take place and these higher vibrational levels form the lowest compound state of O_2. A great deal of attention has been given to the determination of the energies of the vibrational states of $O_2^-(X^2\Pi_g)$. The electron affinity of O_2 is reliably known to be 0.440 ± 0.008 eV (Celotta et al., 1972). The level $v' = 4$ is the first level that lies above the ground state of O_2 (and can thus be classified as a resonance). Spin-orbit coupling splits this level into two components (Land and Raith, 1974) separated by 20 ± 2 meV. Using a time-of-flight transmission apparatus, Land and Raith find the energy of the *center* of this doublet at 91 ± 5 meV. This value should be compared with the value of 76 meV from the "raw data" of Linder and Schmidt (1971b), which they correct to 82 meV to take into account the rotational structure in the resonance. In view of the doublet structure of O_2^-, which was not known to Linder and Schmidt, this correction is probably in error (Land and Raith, 1974).

An additional tie point in the vibrational ladders of O_2^- and O_2 is provided by the observation of Spence and Schulz (1970) that an accidental coincidence in energy occurs (within ~10 meV) between the $v = 3$ level of $O_2(X^3\Sigma_g^-)$ at 570 meV and the $v' = 8$ level of $O_2^-(X^2\Pi_g)$.

The vibrational levels $v' > 4$ provide the resonances through which vibrational excitation occurs. Because the sequence of spikes forming the vibrational cross section is so sharp for O_2, it is best to follow the suggestion of Linder and Schmidt (1971) and present the energy-integrated cross sections. These are shown in the form of a histogram in Fig. 1.29, spanning the range from 0.33 to 1.65 eV, corresponding to the vibrational quantum numbers of O_2^- from $v' = 6$ to $v' = 18$. The values obtained by Linder and Schmidt (1971), together with the widths of resonance spikes obtained by Koike and Watanabe (1973), are shown in Table 1.2. Although the values quoted in Table 1.2 are larger than previously measured cross sections (Spence and Schulz, 1970), an analysis of transport coefficients by Phelps (private communication) using resonance scattering only shows that these cross sections are still too low and that the values shown in Table 1.2 must be increased by a factor of approximately 2 to bring them into accord with swarm experiments. The width for the $v' = 4$ state is listed as 0.004 meV in Table 1.2. Herzenberg (1969) gives an estimate of 0.002 meV but the confidence limit for the width is probably not better than a factor of 2.

It should be noted that the next higher shape resonance in O_2, which has been identified by Wong et al. (1973) as the $^4\Sigma_u^-$ state, leads to a broad vibrational excitation, without fine structure, in the energy range 6 to 12 eV. The differential cross section for this process is shown in Fig. 1.30.

No detailed analysis has yet been made regarding the role of the direct component of vibrational excitation. Thus we are left for the present time with the resonance component as the only explored mechanism.

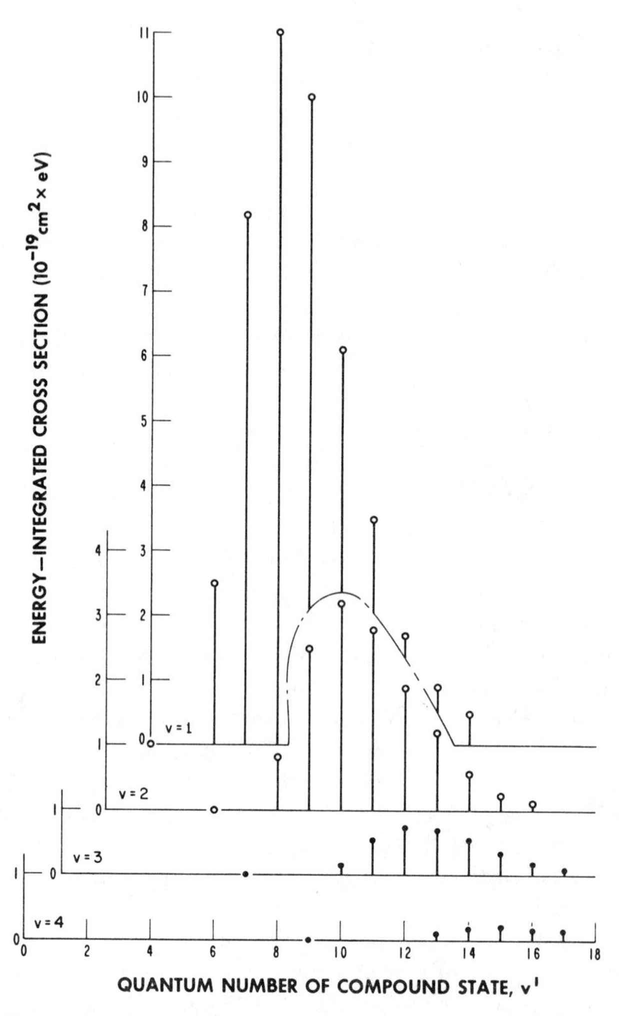

Fig. 1.29 Energy-integrated cross section for vibrational excitation to $v = 1, 2, 3, 4$ versus quantum number of compound state, as measured by Linder and Schmidt (1971) for O_2. The cross sections consist of a series of narrow spikes at the positions of the vibrational levels, v', of the compound state.

Table 1.2 Resonance energies of $O_2^-(X^2\Pi_g)$, widths of $O_2^-(X^2\Pi_g)$, and energy-integrated vibrational cross sections [energies and cross sections from Linder and Schmidt (1971b); widths from Koike and Watanabe (1973)]

v' [a]	4 [h]	5	6	7	8	9	10	11	12	13	14	15	16	17	18
E[eV] [b]	0.082	0.207	0.330	0.450	0.569	0.686	0.801	0.914	1.025	1.135	1.242	1.346	1.449	1.550	1.649
ΔE[meV] [c]		125	123	120	119	117	115	113	111	110	107	104	103	101	99
Γ[meV] [d]	0.004	0.036	0.12	0.26	0.46	0.74	1.1								
$\bar{Q}_v$ (in 10^{-20} cm^2 × eV) [e]															
$v=1$ [f]	x [g]	—	25	82	110	100	61	35	17	9	5	—	—	—	—
$v=2$	x	x	x	—	8.5	25	32	28	19	12	5.8	2.4	1.0	—	—
$v=3$	x	x	x	x	—	—	1.3	5.5	7.3	7.0	5.8	3.3	1.8	1.0	—
$v=4$	x	x	x	x	x	x	—	—	—	1.0	1.9	2.0	1.7	1.0	1.0

[a] v' is the vibrational quantum number of the $O_2^-(X^2\Pi_g)$ compound state.
[b] E is the energy of the compound state in the v' level (in eV).
[c] ΔE is the spacing of the vibrational levels (in meV) of the compound state.
[d] Γ is the width of the vibrational level v' (in meV).
[e] $\bar{Q}_v$ is the energy-integrated cross section to the final vibrational state v of $O_2(X^3\Sigma_g^-)$, which proceeds via $O_2^-(v')$ [in units of 10^{-20} cm^2 × eV].
[f] v is the quantum number of vibrational states of $O_2(X^3\Sigma_g^-)$.
[g] The symbol x indicates that the state is energetically inaccessible.
[h] Using a time-of-flight spectrometer, Land and Raith (1974) locate the center of the $v'=4$ state at 0.091 ± 0.005 eV, i.e., 8 meV above the value given in this table. The splitting of the $v'=4$ state of $O_2^-(X^2\Pi_g)$ is $\Delta E = 20 \pm 2$ meV.

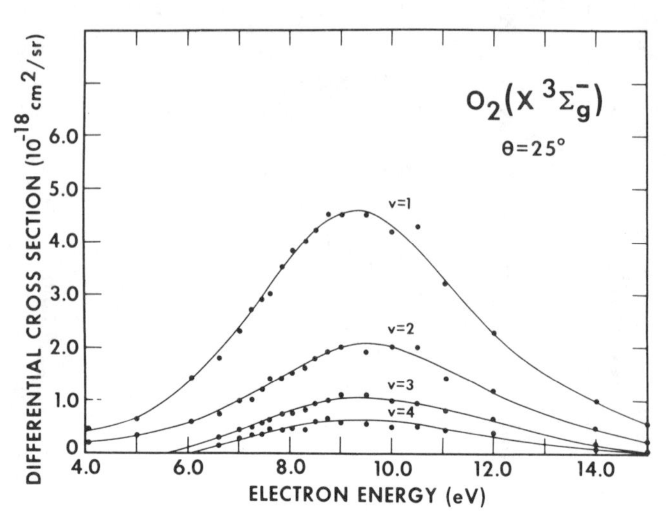

Fig. 1.30 Vibrational excitation of O_2 in the energy range 4–14 eV. (From Wong and Schulz, 1973.)

5.2 NO

The lowest negative-ion state in NO has a configuration $^3\Sigma^-$ and its electron affinity is 24(+10, −5) meV (Siegel et al., 1972). Higher members of the same electronic configuration have symmetries $^1\Delta$ and $^1\Sigma^+$, respectively (Burrow, 1974). Burrow (1974) finds that an abrupt change occurs in the lineshape and peak spacing when the elastically scattered current at 180° is examined. This change, near an impact energy of 0.7 eV, is associated with the threshold of the $^1\Delta$ state. Whereas the spacing of the vibrational levels of the $^3\Sigma^-$ state is 165 meV, the spacing of the levels of the $^1\Delta$ state is 182 meV. Previous experiments in NO have also observed structure in this energy range (Boness and Hasted, 1966; Boness, Hasted, and Larkin, 1968; Spence and Schulz, 1970; Ehrhardt and Willmann, 1967; Zecca et al., 1974), but they have attributed the structure to a single electronic state of NO^-. A mild inconsistency exists here, since one would expect that Zecca et al. (1974), who measured the total cross section, would have also observed the effect of the $^1\Delta$ state, but they did not choose to interpret their data in this manner. Tronc et al. (1975) find that the third member of the multiplet $^1\Sigma^+$ must be taken into account starting at 1.12 eV. They find a value of ω_e =169 meV for $^3\Sigma^-$ ($\omega_e x_e$ =1 meV) and ω_e =185 meV for $^1\Delta$, in agreement with the results of Burrow (1974). Spence and Schulz (1970) find a value of ω_e =170 meV ($\omega_e x_e$ =1 meV) for $^3\Sigma^-$.

Figure 1.31 shows the vibrational excitation functions to v =1, 2, 3, 4, 5

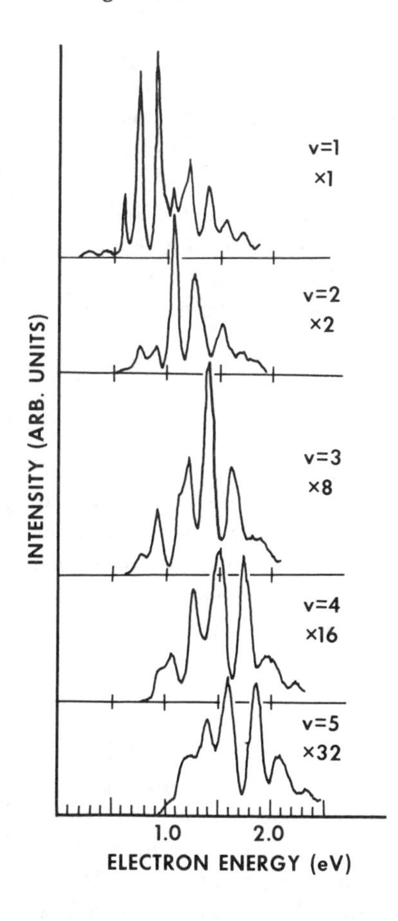

Fig. 1.31 Energy dependence of the differential cross section for vibrational excitation to $v = 1$ to 5 in NO, at a scattering angle of 40°. The numbers on the right give the signal amplification with respect to the $v = 1$ spectrum. (From Tronc et al., 1975.)

as observed by Tronc et al. (1975) at an angle of 40°. These authors do not observe a direct component for vibrational excitation, which one may expect, as a result of the dipole moment. Because three electronic states of the negative ion exist in the range of interest (up to 2.5 eV), NO is a much more complex case than has been anticipated. For example, Tronc et al. (1975) find that the two lowest resonances ($^3\Sigma^-$ and $^1\Delta$) can be classified as long-lived states, whereas the third resonance, $^1\Sigma^+$, is probably a boomerang state. Of course interference can be expected between these states in some energy ranges.

Consideration of the branching ratio indicates that the barrier height involved in the case of NO is lower than that in O_2 and that probably a p-wave dominates the electron escape. The resulting lower barrier leads to a shorter lifetime of NO^- compared to the lifetime of O_2^- (about 10^{-10} sec). The assumption of a p-wave barrier is consistent with the theoretical considerations of Bardsley and Read (1968), who point out that the partial waves being mixed are the $p\pi$ and $d\pi$ waves. Bardsley and Read also point out that in resonance formation and decay at low energies, a p-wave component is much more efficient than a d-wave component, since the centrifugal barrier through which the incoming or outgoing electron must tunnel is much lower for p-waves than for d-waves.

According to Spence and Schulz (1971), who normalized their trapped-electron current to the positive-ion current, the absolute magnitude of the low-lying peaks (0.27 eV, 0.43 eV) is about 7×10^{-18} cm^2. But one must await further experiments before the absolute magnitude can be reliably known.

6 OTHER MOLECULES

In this section we discuss the vibrational excitation of those molecules of which our knowledge is still fragmentary. Often we know the location and widths of resonances rather well, but the decay modes into particular vibrational levels may not have been studied in sufficient detail. Still some estimates can be made regarding vibrational excitation.

6.1 Benzene (C_6H_6)

The first two shape resonances in benzene, which result from the addition of an electron to the first empty π-orbital, are degenerate. The spacing of the $C_6H_6^-$ modes as determined from electron transmission experiments (Sanche and Schulz, 1973) indicates that the temporary negative ion vibrates in the completely symmetric breathing mode. In both C_6H_6 and in C_6D_6 the structures are sharp, with a width of about 0.08 eV, indicating a long-lived compound state. Vibrational excitation of C_6H_6 or C_6D_6 in this energy range (1 to 1.6 eV) has not yet been studied in detail and one would expect, on the basis of the long-lived compound-state model, that a series of spikes would be present in the vibrational excitation, at the energies of the resonances. The experiment of Larkin and Hasted (1972) in which forward-scattered electrons are analyzed in energy actually shows a number of resolved peaks in the vibrational cross sections, thus confirming the preceding expectation (Hasted, 1973). A third shape resonance with a center at 4.8 eV and a width of about 1 eV has been observed in transmission experiments. The decay of this resonance into the ν_1 vibrational mode has been studied by Azria and Schulz (1975) and its decay into the optically allowed $^1B_{2u}$ state has been observed by Smyth, Schiavone, and Freund (1974). The absolute magnitude of the vibrational cross sections is not known.

6.2 Hydrocarbons

The total scattering cross section in methane exhibits a remarkable similarity to argon, especially in the neighborhood of the Ramsauer minimum as has been pointed out by Massey (1969). Other saturated hydrocarbons—namely ethane (C_2H_6), propane (C_3H_8), and butane (C_4H_{10})—also exhibit a Ramsauer minimum (Massey, 1969; Duncan and Walker, 1974). However cyclopropane does not exhibit such a Ramsauer minimum (Duncan and Walker, 1974).

The vibrational cross sections at low energies (0.1 to 1 eV) of methane (Cottrell and Walker, 1965; Pollock, 1968; Duncan and Walker, 1972a) as well as the other hydrocarbon molecules are in the 10^{-16} cm^2 range, as deduced from an analysis of swarm experiments (Duncan and Walker, 1974). The results of such evaluations are shown in Figs. 1.32 to 1.34. The physical mechanisms leading to the large vibrational cross sections in these simple hydrocarbons are not established. No relationship has been developed so far between Ramsauer

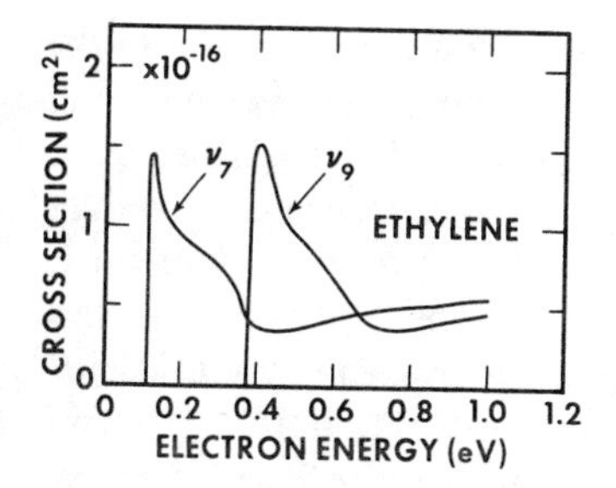

Fig. 1.32 Vibrational cross section for ethylene derived from electron transport coefficients under the assumption that the ν_7 and ν_9 modes are excited. (From Duncan and Walker, 1972b.)

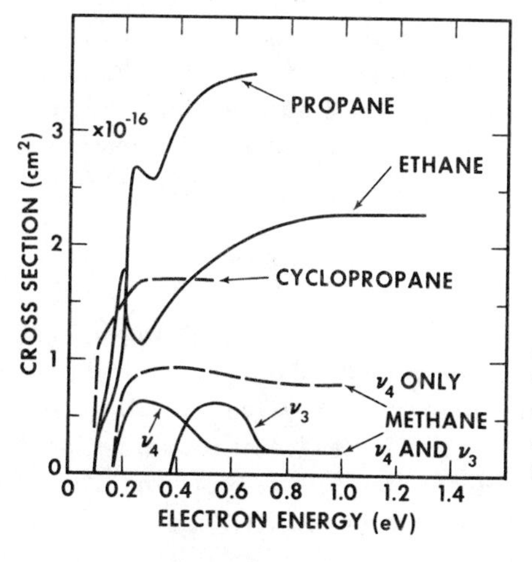

Fig. 1.33 Vibrational cross sections for hydrocarbons, as derived from transport coefficients by Duncan and Walker (1972, 1974). For methane, two possibilities are shown: The dashed line assumes that only the ν_4 mode is excited; the solid lines are calculated under the assumption that the ν_3 mode as well as the ν_4 mode are excited. (From Duncan and Walker, 1972, 1974.)

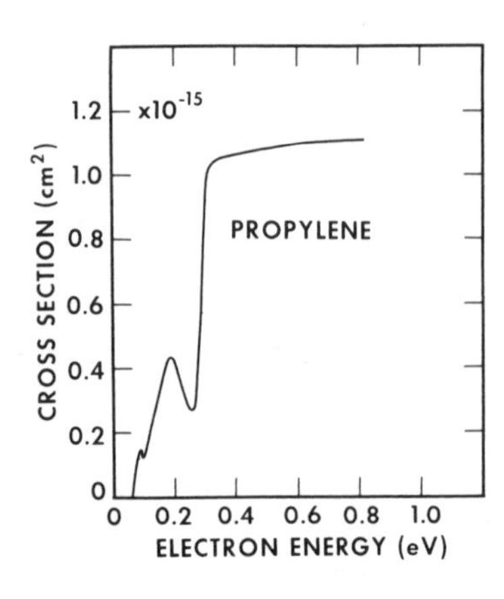

Fig. 1.34 Vibrational excitation of propylene derived from transport coefficients. (From Duncan and Walker, 1974.)

minima and resonances, and thus it would appear that independent processes are involved.

In ethylene (C_2H_4) Boness et al. (1967) do find resonance structure in the vicinity of 0.2 eV, which is not understood. In methane (CH_4) they only find a broad peak near 2.4 eV with some weak fine structure that could be due to an N_2 impurity. Bowman and Miller (1965) and also Dance and Walker (1973) do find trapped-electron peaks, indicating the presence of inelastic processes in the 1.7 to 2 eV range in ethylene, propylene, and acetylene, but these energies are too high to be characteristic of the low-energy vibrational excitation discussed in connection with Figs. 1.32 to 1.34. Rather the 2 eV range may be characteristic of a resonance in which the extra electron occupies a low-lying π^* orbital. Dance and Walker (1973) also studied several other unsaturated molecules, and in all of these a resonance is observed in the neighborhood of 2 eV. Burrow and Jordan (1975) find that the resonance in ethylene in the energy range 1.5 to 2.3 eV exhibits weak fine structure, spaced about 0.17 eV apart, which may indicate that the resonance is of the boomerang type.

6.3 Hydrogen Halides

The vibrational cross section to the $v = 1$ level in HCl shows two distinct peaks (Rohr and Linder, 1975), a sharp one near threshold and a broad peak in the energy range 2 to 4 eV. This finding is very reminiscent of the vibrational cross section in H_2O, as discussed in Sec. 4.3. It should be noted that both H_2O and HCl have large dipole moments (1.85 D and 1.11 D, respectively). The angular distribution of electrons in HCl is isotropic for both peaks. The trapped-electron experiments of Ziesel et al. (1975) show that the first peak of the vibrational cross section to the $v=1$ level occurs within 60 meV of the threshold.

The absolute magnitude of the peak cross section for $v=1$ near threshold is found to be 1.3×10^{-15} cm^2 $\pm 50\%$ (Rohr and Linder, 1975). This value is within the range given by Ziesel et al. (1975) and the mean value (0.8 to 1.7 eV) determined by Center and Chen (1974) from the infrared emission in a discharge. The vibrational excitation to $v=2$ is about a factor of 10 lower in the threshold region (Ziesel et al., 1975).

The preceding experimental observations, together with the dissociative attachment data of Ziesel et al. (1975), can be interpreted by postulating that two potential energy curves ($^2\Sigma^+$) of HCl^- traverse the Franck-Condon region. The lower of these, $HCl^-(^2\Sigma^+)$, correlated with $Cl^- + H$ at infinity, would then be responsible for the threshold peak in vibrational excitation as well as the formation of Cl^- from HCl, as pointed out by Fiquet-Fayard (1974b). The upper potential energy curve invoked by Wong (private communication) would be correlated with $H^- + Cl$ and would be responsible for the 2.5 eV peak. The interpretation of Rohr and Linder (1975) is on the basis of a single $^2\Sigma^+$ potential energy curve, with a virtual state being responsible for the threshold peak in vibrational excitation. Detailed calculations must be awaited to decide between these two models.

7 CONCLUSIONS

Table 1.3 lists, for selected molecules, the properties of the lowest resonances. The molecules are listed in groupings according to the lifetime of the resonance: First is the group of molecules having resonances of intermediate lifetime ($\tau \cong 10^{-14}$ sec), marked B for *boomerang*; second, the group having resonances of short lifetime ($\tau = 10^{-15} - 10^{-14}$ sec), marked S; third, molecules with resonances of long lifetime ($\tau = 10^{-14} - 10^{-10}$ sec), marked L; and finally a group of unclassified molecules. The dominant partial wave is listed in the third column. The fourth column lists the energy range of the resonance for boomerang states and for long-lived states; for short-lived states the center of the resonance is listed and the range of the resonance can be deduced from the half-width, which is given in the next column.

Also listed in Table 1.3 is the energy at which the first fine structure occurs for long-lived and boomerang resonances. In boomerang molecules the energy of the first fine structure changes with the final decay channel and the value in Table 1.3 is the lowest value reported. The nature and the number of vibrational states of the resonance (XY^-) are listed: For example, in CO_2 one can observe 18 ($n00$) modes.

Table 1.3 Lowest shape resonances in selected molecules

		First shape resonance									Second shape resonance			
Molecule	Designation	Dominant partial wave	Energy range or center (eV)	Half-width (eV)	First fine structure (eV)	Fine structure spacing (eV)	XY^- Modes observed (number)	Approx. no. vibrational levels excited	σ Max (10^{-16} cm^2)	Model	Energy range or center (eV)	Half-width (eV)	σ Max (cm^2)	Designation
N_2	$^2\Pi_g$	$d\pi$	1.7 to 3.5	~0.2	~1.93	~0.3	5	10	3	B	7.5	1		
CO	$^2\Pi$	$d\pi + p\pi$	1.2 to 3.0	~0.4	~1.6	~0.27	5	10	3.5	B				
CO_2	$^2\Pi_u$	$p\pi$	3 to 5	~0.2	~3.14	0.131	n00(18)	25	2	B				
C_2H_4														
Ethylene	B_{2g}	d	1.5 to 2.3		~1.5	~0.17	ν_2(3)		2	B				
H_2	$^2\Sigma_u^+$	$p\sigma$	3	6		No fine structure		4	0.4	S				
N_2O	$^2\Pi$			unknown							2.34	1.1	1×10^{-17}	$^2\Sigma^+$
H_2O	$^2B_2$		7.5	5		No fine structure		3	0.6	S				
H_2S			2.3	1.1		No fine structure				S				
O_2	$^2\Pi_g$	$d\pi$	0.08 to 1.5	4×10^{-6}	0.08		15	4	0.1	L	9.5	4	6×10^{-17}	$^4\Sigma_u^-$
NO	$^3\Sigma^-$	$d\pi + p\pi$	0 to 1	<0.05	0.27	0.17	6	5	0.1	L	0.7–			$^1\Delta$
NO_2	$^1A_1$ or $^3B_1$	p	0 to 1	<0.02	0.14	0.13	n00(6)				1 to 2	0.1		
						0.065	0n0(14)							
SO_2	$^2B_1$	p	2.8 to 3.8	0.04	2.8	0.093	n00(10)				3.8 to 6			
C_6H_6	$^2E_{2u}$	f	1 to 1.6	<0.08	1.14	0.123	ν_2(5)				5	1		

In SO_2 there are undoubtedly one or more resonances below the lowest value listed in Table 1.3 (i.e., below 2.8 eV), since SO_2 has a positive electron affinity. The lowest-lying resonances are not detected in transmission experiments (Sanche et al., 1973) or in photodetachment experiments (Feldmann, 1970), probably because of weak coupling (Fiquet-Fayard 1974a).

8 ACKNOWLEDGMENTS

A great contribution to the writing of this chapter was made by A. V. Phelps of JILA, who not only pioneered many of the approaches described here but wrote an early review on the subject and made many suggestions while the author visited JILA. Thanks are also due to P. D. Burrow, A. Herzenberg, K. D. Jordan, W. C. Tam, and S. F. Wong of Yale, who read the manuscript and made numerous suggestions. Appreciation is also expressed to Jean Gallagher of the JILA data center for translating data into digital form. To Lorraine Volsky of JILA and Marion Harrison of Yale go thanks for their struggle with the manuscript.

Support for portions of the work described in the text was received from the National Science Foundation, the Office of Naval Research, and the Army Research Office Durham, and this support is gratefully acknowledged.

REFERENCES

Abrams, R. A. and A. Herzenberg, 1969, "Rotational Excitation of H_2 by Slow Electrons," *Chem. Phys. Lett.* **3**, 187.

Andrick, A., D. Danner, and H. Ehrhardt, 1969, "Vibrational Excitation of CO_2 by Dipole Interaction with Slow Electrons," *Phys. Lett.*, **A29**, 346.

Andrick, A. and F. H. Read, 1971, "Angular Distributions for the Excitation of Vibronic States by Resonant Electron Molecule Reactions," *J. Phys.*, **B4**, 389.

Azria, R. and G. J. Schulz, 1975, "Vibrational and Triplet Excitation by Electron Impact in Benzene," *J. Chem. Phys.*, **62**, 573.

Azria, R., S. F. Wong, and G. J. Schulz, 1975, "Vibrational Excitation in N_2O via the 2.3 eV Shape Resonance," *Phys. Rev.*, **A11**, 1309.

Bardsley, J. N., A. Herzenberg, and F. Mandl, 1966, "Electron Resonances of the H_2^- Ion," *Proc. Phys. Soc. (London)*, **89**, 305.

Bardsley, J. N. and F. H. Read, 1968, "Predicted Angular Distribution for Resonant Scattering of Electrons by Molecules," *Chem. Phys. Lett.* **2**, 333.

Benedict, W. S., M. A. Pollack, and W. J. Tomlinson, 1969, "The Water-Vapor Laser," *IEEE J. Quantum Electron.*, **5**, 108.

Birtwistle, D. T. and A. Herzenberg, 1971, "Vibrational Excitation of N_2 by Resonance Scattering of Electrons," *J. Phys.*, **B4**, 53.

Blatt, J. M. and V. F. Weisskopf, 1952, *Theoretical Nuclear Physics*, Wiley, New York.

Boness, M. J. W. and J. B. Hasted, 1966, "Resonances in Electron Scattering by Molecules," *Phys. Lett.*, **21**, 526.

Boness, M. J. W., I. W. Larkin, J. B. Hasted, and L. Moore, 1967, "Virtual Negative Ion Spectra of Hydrocarbons," *Chem. Phys. Lett.*, **1**, 292.

Boness, M. J. W., J. B. Hasted, and I. W. Larkin, 1968, "Compound State Electron Spectra of Simple Molecules," *Proc. Roy. Soc. (London)*, **A305**, 493.

Boness, M. J. W. and G. J. Schulz, 1973, "Excitation of High Vibrational States of N_2 and CO via Shape Resonances," *Phys. Rev.*, **A8**, 2883.

Boness, M. J. W. and G. J. Schulz, 1974, "Vibrational Excitation in CO_2 via the 3.8 eV Resonance," *Phys. Rev.*, **A9**, 1969.

Bowman, C. R. and W. D. Miller, 1965, "Excitation of Methane, Ethane, Ethylene, Propylene, Acetylene, Propyne and 1-Butyne by Low-Energy Electron Beams," *J. Chem. Phys.*, **42**, 681.

Breig, E. L. and C. C. Lin, 1965, "Vibrational Excitation of Diatomic Molecules by Electron Impact," *J. Chem. Phys.*, **43**, 3839.

Brown, S. C., 1959, *Basic Data of Plasma Physics*, Technology Press, Wiley, New York.

Burrow, P. D., 1974, "Temporary Negative Ion Formation in NO and O_2," *Chem. Phys. Lett.* **26**, 265.

Burrow, P. D. and G. J. Schulz, 1969, "Vibrational Excitation by Electron Impact Near Threshold in H_2, D_2, N_2 and CO," *Phys. Rev.*, **187**, 97.

Burrow, P. D. and K. D. Jordan, 1975, "On the Electron Affinities of Ethylene and 1,3-Butadiene," *Chem. Phys. Lett.*, **36**, 594.

Celotta, R. J., R. A. Bennett, J. L. Hall, M. W. Siegel, and J. Levine, 1972, "Molecular Photodetachment Spectrometry. II. The Electron Affinity of O_2 and the Structure of O_2^-," *Phys. Rev.*, **A6**, 631.

Center, R. E. and H. L. Chen, 1974, "Vibrational Excitation of HCl by Electron Impact," *J. Chem. Phys.*, **61**, 3785.

Chandra, N. and A. Temkin, 1976, "Hybrid Theory and Calculation of e–N_2 Scattering," *Phys. Rev.*, **A13**, 188.

Chang, E. S., 1974, "Comment on Rotational and Vibrational Excitation of H_2 by Electron Impact," *Phys. Rev. Lett.*, **33**, 1644.

Chantry, P. J., 1969, "Temperature Dependence of Dissociative Attachment in N_2O," *J. Chem. Phys.*, **51**, 3369.

Chen, J. C. Y., 1964, "Theory of Subexcitation Electron Scattering by Molecules," *J. Chem. Phys.*, **40**, 3507.

Chen, J. C. Y., 1966, "Interpretation of the Cross Section for Vibrational Excitation of Molecules by Electrons," *J. Chem. Phys.*, **45**, 2710.

Christophorou, L. G., 1971, *Atomic and Molecular Radiation Physics*, Wiley-Interscience, New York.

Claydon, C. R., G. A. Segal, and H. S. Taylor, 1970, "Theoretical Interpretation of the Electron Scattering Spectrum of CO_2," *J. Chem. Phys.*, **52**, 3387. See also M. Krauss and D. Neumann.

Cottrell, T. L. and I. C. Walker, 1965, "Drift Velocities of Slow Electrons in Polyatomic Gases," *Trans. Faraday Soc.*, **61**, 1585.

Crompton, R. W., M. T. Elford, and A. I. McIntosh, 1968, "Electron Transport Coefficients in Hydrogen and Deuterium," *Aust. J. Phys.*, **21**, 43.

Dance, D. F. and I. C. Walker, 1973, "Threshold Electron Energy-Loss Spectra for Some Unsaturated Molecules," *Proc. Roy. Soc. (London)*, **A334**, 259.

Danner, D., 1970, Thesis, University of Freiburg, unpublished.

Dubé, L. and A. Herzenberg, 1975, "Resonant Electron-Molecule Scattering: The Impulse Approximation in N_2O," *Phys. Rev.*, **A11**, 1314.

Duncan, C. W. and I. C. Walker, 1972a, "Collision Cross Sections for Low Energy Electrons in Methane," *J. Chem. Soc. (London)*, **68**, 1514.

Duncan, C. W. and I. C. Walker, 1972b, "Collision Cross Sections for Electrons in Ethylene and Acetylene, *J. Chem. Soc. (London)*, **68**, 1800.

Duncan, C. W. and I. C. Walker, 1974, "Collision Cross Section for Low-Energy Electrons in Some Simple Hydrocarbons," *J. Chem. Soc. (London)*, **70**, 577.

Ehrhardt, H., L. Langhans, F. Linder, and H. S. Taylor, 1968, "Resonance Scattering of Slow Electrons from H_2 and CO Angular Distributions," *Phys. Rev.*, **173**, 222.

Ehrhardt, H. and K. Willmann, 1967, "Die Winkelabhängigkeit der Resonanzstreuung Niederenergetischer Elektronen an N_2," *Z. Phys.*, **204**, 462.

Engelhardt, A. G., A. V. Phelps, and C. R. Risk, 1964, "Determination of Momentum Transfer and Inelastic Cross Sections for Electrons in Nitrogen Using Transport Coefficients," *Phys. Rev.*, **135**, A1566.

Feldman, D., 1970, "Photoablösung Von Elektronen Bei Einigen Stabilen Negativen Ionen," *Z. Naturforsch*, **25a**, 621.

Fiquet-Fayard, F., 1974a, "Theoretical Problems in the Interpretation of Dissociative Attachment Experiments," *Vacuum*, **24**, 533.

Fiquet-Fayard, F., 1974b, "Theoretical Investigation of Dissociative Attachment in HCl and DCl," *J. Phys.*, **B7**, 810.

Gilardini, A. L., 1972, *Low Energy Electron Collisions in Gases*, Wiley, New York.

Hake, R. D., Jr., and A. V. Phelps, 1967, "Momentum-Transfer and Inelastic-Collision Cross Sections for Electrons in O_2, CO and CO_2," *Phys. Rev.*, **158**, 70.

Hasted, J. B., 1964, *Physics of Atomic Collisions*, Butterworth, London.

Hasted, J. B., 1973, "Electron Scattering Spectroscopy," *Contemp. Phys.*, **14**, 357.

Henry, R. J. W. and E. S. Chang, 1972, "Rotational-Vibrational Excitation of H_2 by Slow Electrons," *Phys. Rev.*, **A5**, 276.

Herzberg, G., 1950, *The Spectra of Diatomic Molecules*, Van Nostrand Reinhold, New York.

Herzenberg, A., 1968, "Oscillatory Energy Dependence of Resonant Electron-Molecule Scattering," *J. Phys.*, **B1**, 548.

Herzenberg, A., 1969, "Attachment of Slow Electrons to Oxygen Molecules," *J. Chem. Phys.*, **51**, 4942.

Herzenberg, A. and F. Mandl, 1962, "Vibrational Excitation of Molecules by Resonant Scattering of Electrons," *Proc. Roy. Soc. (London)*, **A270**, 48.

Hughes, B. M., C. Lifshitz, and T. O. Tiernan, 1973, "Electron Affinities from Endothermic Negative-Ion Charge-Transfer Reactions. III. NO, NO_2, SO_2, CS_2, Cl_2, Br_2, I_2 and C_2H," *J. Chem. Phys.*, **59**, 3162.

Itikawa, Y., 1970, "Differential Cross Section for Vibrational Excitation of CO by Slow-Electron Collision," *J. Phys. Soc. Japan*, **28**, 1062.

Itikawa, Y., 1971, "Nonresonant Vibrational Excitation of CO_2 by Electron Collision," *Phys. Rev.*, **A3**, 831.

Itikawa, Y., 1974, "Electron-Impact Vibrational Excitation of H_2O," *J. Phys. Soc. Japan*, **36**, 1127.

Itikawa, Y. and K. Takayanagi, 1969, "Vibrational Excitation of CO by Slow-Electron Collision," *J. Phys. Soc. Japan*, **27**, 1293.

Javan, A., W. R. Bennett, Jr., and D. R. Herriott, 1961, "Population Inversion and Continuous Optical Maser Oscillation in a Gas Discharge Containing the He–Ne Mixture," *Phys. Rev. Lett.*, **6**, 106.

Jeffers, W. Q. and C. E. Wiswall, 1971, "Excitation and Relaxation in a High Pressure CO Laser," *J. Quant. Electron*, **QE-7**, 407.

Kieffer, L. J., 1973, "A Compilation of Electron Collision Cross Section Data for Modelling Gas Discharge Lasers," JILA Inf. Ctr. Rep. 13.

Koike, F. and T. Watanabe, 1973, "On the Mechanism of Electron Attachment by O_2," *J. Phys. Soc. Japan*, **34**, 1022.

Krauss, M. and F. H. Mies, 1970, "Molecular-Orbital Calculation of the Shape Resonance in N_2," *Phys. Rev.* **A1**, 1592.

Krauss, M. and D. Neumann, 1972, "Energy Curves of CO_2^-," *Chem. Phys. Lett.*, **14**, 26.

Land, J. E. and W. Raith, 1974, "High-Resolution Measurement of Resonances in e–O_2 Scattering by Electron Time-of-Flight Spectroscopy," *Phys. Rev.*, **A9**, 1592.

Larkin, I. W. and J. B. Hasted, 1972, "Electron Transmission Studies of Decay Channels of Molecular Resonances," *J. Phys.*, **B5**, 95.

Linder, F. and H. Schmidt, 1971a, "Rotational and Vibrational Excitation of H_2 by Slow Electron Impact," *Z. Naturforsch.*, **26a**, 1603.

Linder, F. and H. Schmidt, 1971b, "Experimental Study of Low Energy e–O_2 Collision Processes," *Z. Naturforsch.*, **26a**, 1617.

Lowke, J. J., A. V. Phelps, and B. W. Irwin, 1973, "Predicted Electron Transport Coefficients and Operating Characteristics of CO_2–N_2–He Laser Mixtures," *J. Appl. Phys.*, **44**, 4664.

Massey, H. S. W., 1969, *Electronic and Ionic Impact Phenomena*, Clarendon, Oxford.

McDaniel, E. W., 1964, *Collision Phenomena in Ionized Gases*, Wiley, New York.

Nighan, W. L., 1970, Electron Energy Distributions and Collision Rates in Electrically Excited N_2, CO and CO_2, *Phys. Rev.*, **A2**, 1989.

Patel, C. K. N., 1964, "Selective Excitation through Vibrational Energy Transfer and Optical Maser Action in N_2–CO_2," *Phys. Rev. Lett.*, **13**, 617.

Patel, C. K. N., 1965, "CW Laser Action in N_2O (N_2–N_2O System)," *Appl. Phys. Lett.*, **6**, 12.

Patel, C. K. N., 1968, *Lasers*, A. K. Levine (Ed.), Marcel Dekker, New York.

Pavlovic, Z., M. J. W. Boness, A. Herzenberg, and G. J. Schulz, 1972, "Vibrational Excitation in N_2 by Electron Impact in the 15–35 eV Region," *Phys. Rev.*, **A6**, 676.

Penner, S. S., 1959, *Quantitative Molecular Spectroscopy and Gas Emissivities*, Addison-Wesley, Reading, Mass.

Phelps, A. V., 1968, "Rotational and Vibrational Excitation of Molecules by Low-Energy Electrons," *Rev. Mod. Phys.*, **40**, 399.

Pollock, W. J., 1968, "Momentum Transfer and Vibrational Cross Sections in Non-Polar Gases," *Trans. Faraday Soc.*, **64**, 2919.

Rohr, K. and F. Linder, 1975, "Vibrational Excitation in e–HCl Collisions at Low Energies," *J. Phys.*, **B8**, L200.

Sanche, L. and G. J. Schulz, 1973, "Electron Transmission Spectroscopy: Resonances in Triatomic Molecules and Hydrocarbons," *J. Chem. Phys.* **58**, 479.

Schulz, G. J., 1964, "Vibrational Excitation of N_2, CO and H_2 by Electron Impact," *Phys. Rev.*, **135**, A988.

Schulz, G. J., 1973, "Resonances in Electron Impact on Atoms," *Rev. Mod. Phys.*, **45**, 378.

Schulz, G. J., 1973, "Resonances in Electron Impact on Diatomic Molecules," *Rev. Mod. Phys.*, **45**, 423.

Seng, G. and F. Linder, 1974, "Scattering Mechanisms in Low-Energy e–H_2O Collisions," *J. Phys.*, **B7**, L509.

Siegel, M. W., R. J. Celotta, J. L. Hall, J. Levine, and R. A. Bennett, 1972, "Molecular Photodetachment Spectrometry. I. The Electron Affinity of Nitric Oxide and the Molecular Constants of NO^-," *Phys. Rev.*, **A6**, 607.

Smyth, K. C., J. A. Schiavone, and R. S. Freund, 1974, "Electron Impact Excitation of Fluorescence in Benzene, Toluene and Aniline," *J. Chem. Phys.*, **61**, 1782.

Spence, D. and G. J. Schulz, 1970, "Vibrational Excitation by Electron Impact in O_2," *Phys. Rev.*, **A2**, 1802.

Spence, D. and G. J. Schulz, 1971, "Vibrational Excitation and Compound States in NO," *Phys. Rev.*, **A3**, 1968.

Spence, D., J. L. Mauer, and G. J. Schulz, 1972, "Measurement of Total Inelastic Cross Sections for Electron Impact in N_2 and CO_2," *J. Chem. Phys.*, **57**, 5516.

Stamatovic, A. and G. J. Schulz, 1969, "Excitation of Vibrational Modes Near Threshold in CO_2 and N_2O," *Phys. Rev.*, **188**, 213.

Takayanagi, K., 1965, "Vibrational Excitation of Hydrogen Molecule by Slow Electrons," *J. Phys. Soc. Japan*, **20**, 562.

Takayanagi, K., 1965, "Excitation of Molecular Vibration by Slow Electrons," *J. Phys. Soc. Japan*," **20**, 2297.

Takayanagi, K., 1966, "Rotational and Vibrational Excitation of Polar Molecules by Slow Electrons," *J. Phys. Soc. Japan*, **21**, 507.

Takayanagi, K., 1967, "Scattering of Slow Electrons by Molecules," *Progr. Theor. Phys. Suppl. Japan*, **40**, 216.

Tice, R. and D. Kivelson, 1967, "Cyclotron Resonance in Gases. II. Cross Sections for Dipolar Gases and for CO_2," *J. Chem. Phys.*, **46**, 4748.

Temkin, A. and E. C. Sullivan, 1974, "Rotational-Vibrational Coupling in the Theory of Electron-Molecule Scattering," *Phys. Rev. Lett.*, **33**, 1057.

Tien, P. K., D. MacNair, and H. L. Hodges, 1964, "Electron Beam Excitation of Gas Laser Transitions and Measurements of Cross Sections of Excitation," *Phys. Rev. Lett.*, **12**, 30.

Tronc, M., A. Huetz, M. Landau, F. Pichon, and J. Reinhardt, 1975, "Resonant Vibrational Excitation of the NO Ground State by Electron Impact in 0.1–3 eV Energy Range," *J. Phys.*, **B8**, 1160.

Wong, S. F. and G. J. Schulz, 1973, "Vibrational Excitation of O_2 by Electron Impact above 4 eV," *Phys. Rev. Lett.*, **31**, 969.

Wong, S. F. and G. J. Schulz, 1974a, "Rotational and Vibrational Excitation of H_2 by Electron Impact at 4.5 eV: Angular Distribution," *Phys. Rev. Lett.*, **32**, 1089.

Wong, S. F. and G. J. Schulz, 1974b, "Electron Impact Near the $a^3\Pi$ State of CO: Dipole-Dominated Resonances," *Phys. Rev. Lett.*, **33**, 134.

Wong, S. F. and G. J. Schulz, 1975, "Vibrational Excitation of H_2O by Electron Impact at Low Energies," *Abstracts 9th International Conference on the Physics of Electronic and Atomic Collisions*, Seattle, p. 283.

Zecca, A., I. Lazzizzera, M. Krauss, and C. E. Kuyatt, 1974, "Electron Scattering from NO and N_2O below 10 eV," *J. Chem. Phys.*, 61, 4560.

Ziesel, J. P., I. Nenner, and G. J. Schulz, 1975, "Negative Ion Formation, Vibrational Excitation and Transmission Spectroscopy in Hydrogen Halides," *J. Chem. Phys.*, 63, 1943.

CHAPTER 2

Atomic Processes in Planetary Atmospheres

MANFRED A. BIONDI

Department of Physics and Astronomy
University of Pittsburgh
Pittsburgh, Pennsylvania

This paper is derived from the George J. Schulz Memorial lecture presented at Yale University on October 28, 1977.

1 INTRODUCTION

The purpose of this chapter is to trace developments in certain areas of atomic collision physics from the post-World War II period to the present and to show how newly developed experimental techniques first yielded new insights into the fundamental interactions in electron and ion collision processes and how the studies then evolved to provide basic atomic collision data needed for an understanding of the ionized regions in the upper atmospheres of the planets—especially that of the earth. A further evolution—the application of these techniques to direct observations of these atomic collision processes in the earth's ionosphere from ground-based observatories—is also described. The researches used to illustrate these subjects are those which my collaborators and I have carried out; no attempt is made at a comprehensive review of the many contributions of other scientists.

2 THE EARTH'S ATMOSPHERE

To set the stage for the topic of this paper—atomic processes in planetary atmospheres—let us examine a planetary atmosphere of great importance to us—that of the earth—with the aid of Fig. 2.1. As can be seen from the somewhat whimsical drawing of the ionospheric regions of the earth, there are at least two ways in which one can study the ionosphere from the ground. One can go into the research laboratories and study the basic processes, or one can use optical and radio observatories to actually look at the ionosphere. The ionosphere is a plasma (some call it the world's largest plasma physics laboratory), and that plasma is created when the neutral atmosphere of the earth is bombarded by solar radiation and by energetic particles which precipitate during geomagnetic storms. These fluxes of radiation and particles excite and ionize the neutral gas and distinctive layers are formed, the D-, E-, F_1- and F_2-layers, which have rather different characteristics insofar as the important neutral constituents and ionic species are concerned. The plasma so created is a balance between the various processes that generate charged particles and those that remove them, and it is these processes that are discussed here.

My research began in 1946 with a Ph.D. thesis topic dealing with the study of electron removal processes in a plasma after the external ionization had been removed, a period called the afterglow. To study electron removal from a plasma one requires a good method of measuring the behavior of the electron density in that plasma, and a novel method was found by which the newly developed microwave techniques could be

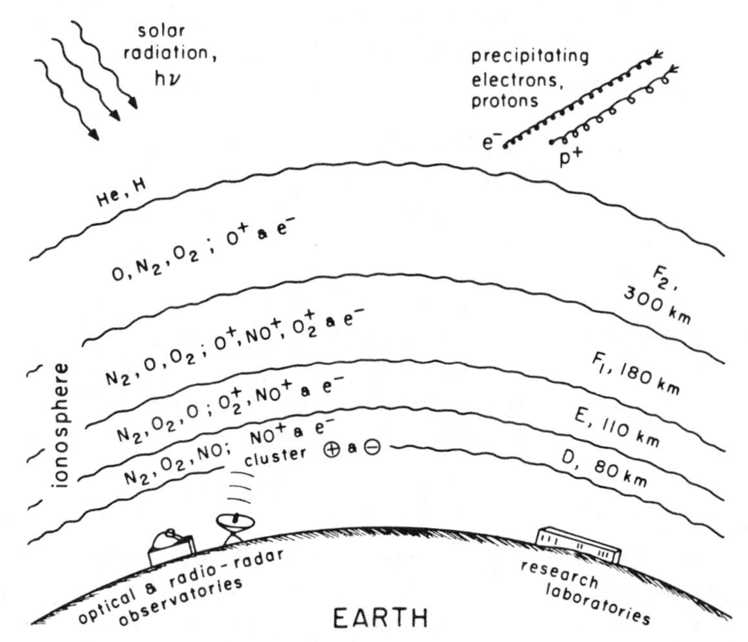

Fig. 2.1 Schematic representation of the various ionospheric regions of the earth's upper atmosphere.

applied to precise determinations of the electron concentration in a partially ionized gas.

A plasma with its free electrons is a dielectric medium, and, if one places such a dielectric between the plates of a capacitor, a change in its capacitance results. If this capacitor is part of a resonant circuit such-as that shown at the top of Fig. 2.2, the resonant frequency of the circuit is changed. The simple equation for the dielectric coefficient of a plasma,

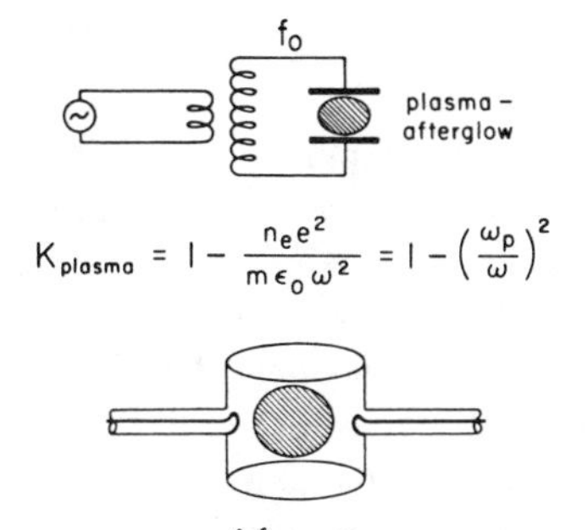

Fig. 2.2 Dielectric effect of a plasma on the resonant frequency: (top) of a tuned circuit; and (bottom) of a resonant microwave cavity.

K_{plasma}, is given in the figure, indicating its dependence on the electron density n_e, as well as on the electronic charge e and mass m, the permittivity ϵ_0 and the angular frequency ω of the microwaves. Alternatively, one can associate a characteristic frequency, the plasma frequency ω_p, with the free electrons in the ionized gas. In actual experiments we do not use a simple tuned circuit such as shown at the top of Fig. 2.2. Instead the plasma is surrounded by a resonant cavity as indicated at the bottom of the figure, yet the resonant frequency of the cavity still changes in proportion to the electron density within the cavity, providing the basis for our method of measurement.

The measurements are taken during the period when the external sources of ionization are removed, the so-called afterglow. The atomic collision processes occurring in the plasma are determined from solutions of the appropriate particle continuity equation, which for electrons is:

$$\frac{\partial n_e}{\partial t} = \sum P_j - \sum L_k - \nabla \cdot \mathbf{\Gamma}_e \tag{2.1}$$

This equation is a statement that if one observes a small volume element of the plasma and studies the behavior of the electrons within that volume, their number density n_e increases due to processes which create new electrons (first term on the right), while n_e decreases as a result of processes which destroy them (second term). Also, n_e decreases if there is a net flow out of that volume (third term). As examples of these processes, electron production may involve a fast electron hitting an atom to knock out an electron, i.e., ionization by electron impact. Alternatively, photons of sufficient energy can photoionize the atom to create an electron-ion pair. Loss terms occur when electrons attach themselves to atoms or molecules which can form stable negative ions. Electrons may also find positive ions and recombine with them to neutralize themselves. Finally, in plasma afterglows the main flow of electrons is by diffusion.

As indicated in (2.2)

$$\frac{\partial n_e}{\partial t} = -(\alpha n_+)n_e - (\beta_{att} N_n)n_e + D_a \nabla^2 n_e \tag{2.2}$$

one can write an equation to describe the afterglow period when the production terms are zero. One finds that the important terms to consider are those that describe the recombination of electrons with ions (recombination coefficient, α), the attachment of the electrons to neutral molecules (coefficient, β_{att}), and ambipolar diffusion (coefficient, D_a). It is difficult experimentally to make sense of such an equation if all these processes occur at the same time, and so one seeks simple conditions. One simple case occurs when the recombination term is the only important one.

Then, as indicated in (2.3),

$$\frac{\partial n_e}{\partial t} \simeq -\alpha n_e^{\,2} \tag{2.3a}$$

yielding

$$\frac{1}{n_e(t)} = \frac{1}{n_e(0)} + \alpha t \tag{2.3b}$$

The equation simplifies, and one has a solution for the electron density variation which is of quite characteristic form. Such solutions of the continuity equation have been used to analyze our measurements.

We first studied the simplest case we could think of, that of helium. The results were more or less what was expected. Helium atoms do not attach electrons to form stable negative ions, and it seemed unlikely that helium ions would recombine very rapidly with electrons. Thus the only thing left was ambipolar diffusion loss, and indeed in helium that was what was found.

Emboldened by success, we moved on to a study of the next noble gas, neon. However when studying neon nothing made sense. The time dependence of the electron decay was not appropriate to diffusion, nor was the variation with the density of neutral neon. We could not understand what was wrong, since electrons cannot attach themselves to neon atoms, and on the basis of theoretical predictions, it was expected that the capture of the electrons by neon atomic ions would be unmeasurably slow.

After puzzling over the data for many months, it finally occurred to me that, in spite of these predictions, the neon data should be compared with the simple recombination decay given in (2.3b). Surprisingly the electron density decay in the afterglow precisely fit a recombination loss; however, the inferred recombination rate was 100 000 times faster than predicted. Thus there was apparently a major discrepancy between theory and experiment.

Fortunately for the progress of physics, at about the same time (1947) in England two British theoretical atomic physicists, Bates and Massey, were concerned about a similar problem in the ionosphere. In the E-region (see Fig. 2.1) it was found that the rapid loss of electrons could not be accounted for by attachment to oxygen molecules. Therefore, they invoked a new process, dissociative recombination, as a possible source of the rapid electron loss (Bates and Massey, 1946, 1947). This process may be written in reaction form as follows:

$$AB^+ + e^- \rightleftharpoons (AB^*)_r \rightarrow A^* + B \tag{2.4}$$

where AB is a hypothetical molecule and the asterisk indicates an excited state. The energy diagram for this reaction is given in Fig. 2.3. As indicated in Fig. 2.1 the ionospheric ions can be molecular in character, and if such a molecular ion is approached by an electron of the appropriate energy ϵ, a resonant capture to form an intermediate excited state can take place (a process akin to the molecular resonances studied by Schulz). Such resonant captures can occur with large probability; however, if nothing further happens the electron is released after a short time. Instead, it is more likely that the repulsive intermediate state of the molecule $(AB^*)_r$ very rapidly breaks up into the atoms A^* and B, thus stabilizing the recombination. On the basis of qualitive estimates, Bates and Massey (1946, 1947) argued that this process might be about 10000 times faster than the fastest recombination process possible when only atomic ions are present in the plasma.

We were unaware of these ionospheric considerations and had no satisfactory theoretical explanation of the findings (since we thought that afterglows contained only atomic ions) when the laboratory results were published in 1949 (Biondi and Brown, 1949a, b). However, upon learning of our results, Bates quickly responded with a letter to Physical Review (Bates, 1950a, b) suggesting that, inasmuch as there was mounting evidence for stable noble gas molecular ions such as He_2^+ and Ne_2^+, the same dissociative recombination process proposed for the ionosphere was operative in our noble gas plasma-afterglows.

By this time I was seeking by experimental studies to learn more concerning the mechanism of this fast recombination process. From Fig. 2.3 it is seen that the atoms formed by dissociative recombination share as kinetic energy the dissociation energy ϵ_D, with the result that the excited atom A^* is moving with a velocity considerably greater than

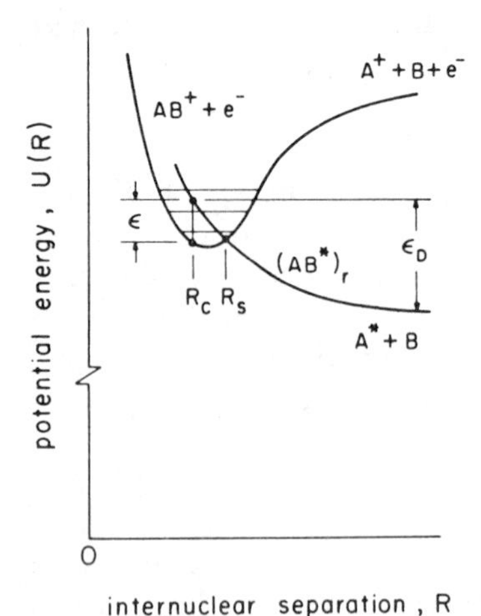

Fig. 2.3 Hypothetical potential energy curves illustrating dissociative recombination of AB^+ ions with electrons.

ordinary thermal velocities. Thus one might hope to see doppler shifts in the radiation emitted if dissociative recombination is indeed the origin of the excited atoms.

Figure 2.4 illustrates what one might expect to observe. Atomic lines are sharply defined in frequency. At absolute zero ($T = 0$) the energy in the line is concentrated very close to the single frequency ν_0; however, at room temperature ($T \sim 300$ K) random thermal motions of the atoms cause doppler shifts that broaden the atomic line slightly (about 1 part in 10^5) and give it the characteristic shape shown at the top of the figure. If these atoms as a group move away from the observer, the line is red-shifted, if towards the observer, blue-shifted. As noted previously, if the dissociative recombination process is operative in a plasma, the dissociation velocities are considerably greater than thermal and are directed in all possible directions, so that an observer should see a rather characteristic, broad line-shape.

Since a plasma-afterglow is an extremely weak source of radiation, in order to see the detailed shape of the recombination line (which is still very narrow by ordinary standards) it was necessary to develop an optical instrument having two usually conflicting properties—very high resolution and very great detection sensitivity. We accomplished this by adapting the high-resolution, multibeam interferometer invented many years ago by Fabry and Perot so that it could use as a signal detector a modern, ultra-sensitive photomultiplier (Biondi, 1956) instead of the usual photographic film (see Fig. 2.4), thereby gaining over a factor of 100 in sensitivity. Although we began studies of the afterglow line shapes in 1951 (Rogers and Biondi, 1964), it was not until 14 years later that it was possible to demonstrate conclusively the dissociative origin of the recombination radiation (Connor and Biondi, 1965; Frommhold and Biondi, 1969).

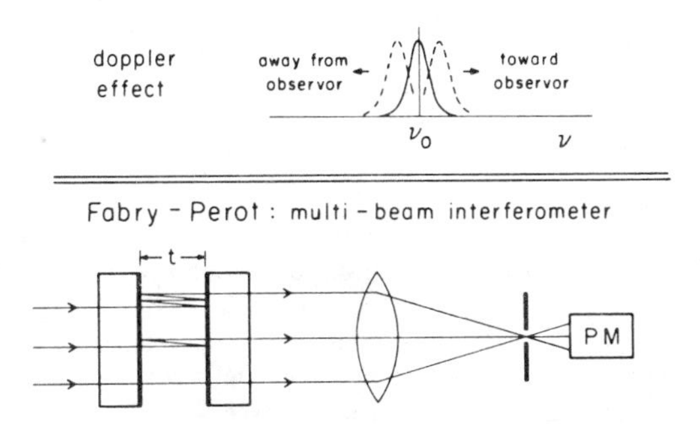

Fig. 2.4 Top: Doppler broadened and shifted spectral line. Bottom: Principle of the Fabry–Perot interferometer adapted for photoelectric detection.

The line shapes observed in neon plasma-afterglows are shown in Fig. 2.5. At the top the narrow line emitted by excited neon atoms moving with near-thermal velocities is seen during the microwave discharge phase of plasma excitation. The thermal broadening of the line profile is sufficiently small that one can detect the isotope shift between the ^{20}Ne (91%) and ^{22}Ne (9%) atoms found in the natural neon samples used. At the bottom of Fig. 2.5 the line profile emitted by the excited atoms produced by recombination during the afterglow is shown. It is seen that in addition to a small thermal core, the profile exhibits the theoretically predicted, broad dissociative "pedestal" indicated in the middle of the figure. Thus, the dissociative nature of the recombination process occurring in noble gas afterglows was demonstrated.

To return to the mid-1950s, our basic studies, which had often centered on the noble gases, now included processes occurring in atmospheric gases such as oxygen and nitrogen. In collaboration with Phelps and Chanin, I was carrying out drift tube studies of the attachment of slow electrons to oxygen molecules (Chanin et al., 1962) while Schulz was using a beam technique to study a similar process at somewhat higher energies. At about this time the Department of Defense discovered that clouds of ionization produced in the

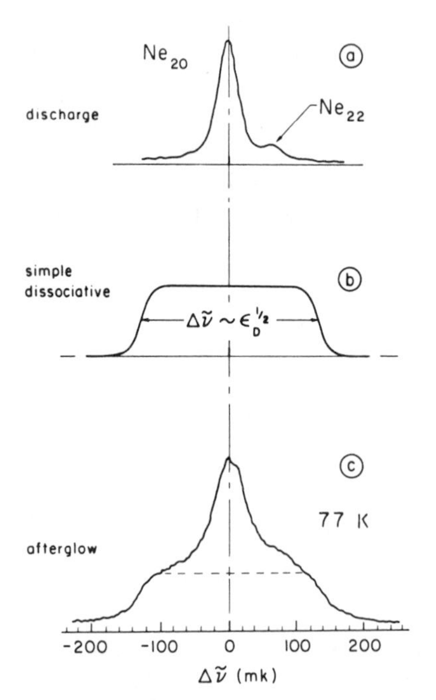

Fig. 2.5 Spectral line shapes of the 5852 Å neon emission. (a) Thermal broadening and isotope shift in the line emitted from the microwave discharge. (b) Predicted line shape for a single dissociation kinetic energy. (c) Dissociatively broadened afterglow line profile.

upper atmosphere by detonation of a nuclear weapon ("nuclear blackout") could blind radar and communications, thus posing a major problem for planned defenses against ballistic missiles. To provide the necessary data for calculations of how the upper atmosphere would respond, laboratory studies of atomic collision processes relating to the earth's ionosphere were given a sudden, major increase in research funding.

One of the critical problems was to obtain quantitative determinations of the recombination coefficients of the important ionospheric ions, which at that time were considered to be O_2^+, NO^+, and N_2^+ (see Fig. 2.1). A difficulty with laboratory studies of these ions is that not only can the desired ion C_2^+ form in oxygen and N_2^+ in nitrogen, but also unwanted ions such as O_3^+, O_4^+, N_3^+, and N_4^+. Thus the microwave apparatus used to determine recombination coefficients from afterglow electron density measurements was modified to include a means for ion identification, as indicated by the schematic diagram of a recent apparatus shown in Fig. 2.6. The resonant cavity provides the means for electron density measurements during the afterglow and also for selective heating of the electrons by microwaves (to reproduce ionospheric conditions). The positive ions in the plasma-afterglow are identified with the aid of the differentially pumped mass-spectrometer.

The measured variation of the recombination coefficient $\alpha(O_2^+)$ with electron temperature (Kasner and Biondi, 1968; Mehr and Biondi, 1969) is shown in Fig. 2.7, which indicates that the recombination rate at ordinary

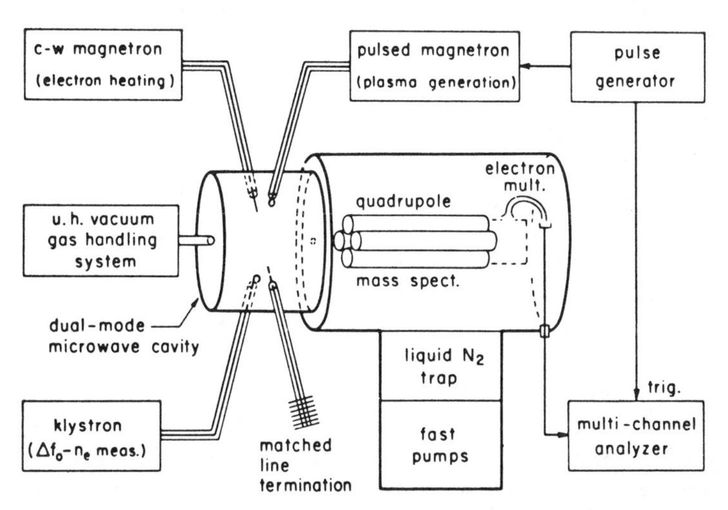

Fig. 2.6 Microwave afterglow apparatus employing controlled heating of the electrons and mass-identification of the ions under study.

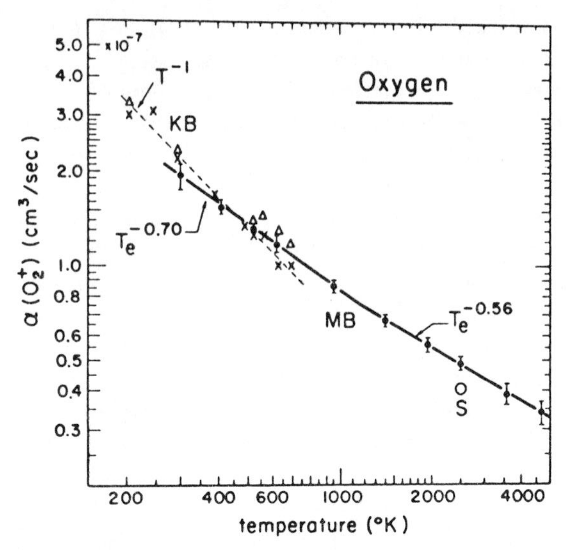

Fig. 2.7 Measured variation of the recombination coefficient for O_2^+ ions and electrons with electron temperature T_e.

temperatures is very large ($\alpha \sim 10^{-7}$ cm^3/sec corresponds to a recombination cross section of 10^{-14} cm^2) and decreases with increasing electron temperature T_e in reasonable accord with theoretical predictions. Studies of this type were carried out for other ionospheric ions, thus helping to provide the data needed for modeling the behavior of the upper atmosphere.

3 OTHER PLANETS

Let us now turn our attention to the other planets, first considering what has been learned about them by observation. From earth-based spectroscopes looking through large telescopes some information has been obtained about the neutral species present in the atmospheres, as suggested in Fig. 2.8. Although these earth-based observations provided information concerning the neutral atmospheres of the planets, it required fly-bys of spacecraft to yield information concerning the structures of the ionospheres.

The planetary-probing spacecraft such as Mariner, Pioneer, and Viking communicate their results back to earth by microwave telemetry links, providing a basis for a "microwave occultation" measurement of the planet's ionospheric electron density profile. As noted earlier in the paper,

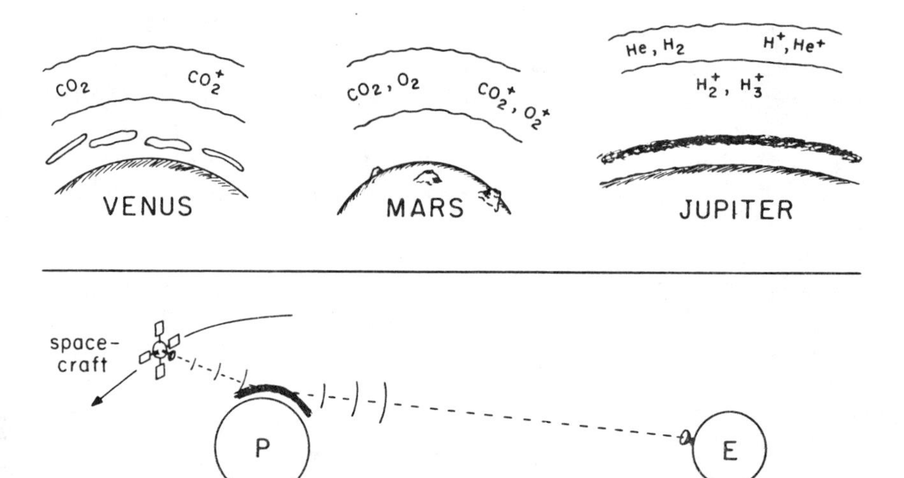

Fig. 2.8 Top: Important neutral and positive ion species in the upper atmospheres of Venus, Mars, and Jupiter (outer regions). Bottom: Microwave occultation experiment during spacecraft flyby of a planet.

electrons in the ionospheric plasma act as a dielectric medium; consequently a microwave signal traversing a region of the ionosphere is refracted (bent) in proportion to the electron concentration distribution along its path (see Fig. 2.8). Since the actual position of the spacecraft as it passes behind the planet can be precisely calculated using orbital mechanics, the bending at each instant is inferred from the difference between the apparent position (given by the microwave tracking station) and the actual position.

The first such experiment was the Mariner fly-by of Mars in the mid-1960s. In order to account for the electron density in the Martian ionosphere deduced from the Mariner occultation experiment, the model makers proposed that photoionization by solar radiation was balanced by electron recombination with CO_2^+ ions (believed to dominate the Martian ionosphere) involving a recombination coefficient which was about 100 times larger than the value we had obtained for $\alpha(O_2^+)$. This value seemed unreasonably large to us; therefore we used our microwave afterglow apparatus to determine $\alpha(CO_2^+)$ under conditions appropriate to the Martian ionosphere and found it to be quite close in value to $\alpha(O_2^+)$, as we expected (Weller and Biondi, 1967). A very lively controversy arose as to who was correct, those making the model calculations or the laboratory investigators. Fortunately for us lab people, a few years later, when a later Mariner was flown past Venus, the ionosphere measured by its occultation experiment was very well reproduced in a model calculation that made use of the laboratory $\alpha(CO_2^+)$ values. From these beginnings in other-planet studies, a NASA-supported research program grew which, of late, has dealt with problems relating to the atmospheres of the outer planets, e.g., Jupiter and Saturn, and to interstellar gas clouds, where ions such as H_3^+, NH_4^+ and HCO^+ are of interest (Leu, Biondi, and Johnsen, 1973a,b; Huang, Biondi, and Johnsen, 1976; Johnsen and Biondi, 1974).

Up to this point the emphasis has been on that part of the ionospheric ionization balance problem involving electron capture by molecular ions. In some regions of planetary ionospheres, *atomic* ions such as H^+, He^+ and O^+ are dominant, and as noted earlier these ions are very inefficient in capturing electrons. Such ions would be long-lived if it were not for ion-molecule reactions.

Of especial interest are those reactions which convert atomic ions to molecular ions (which can then capture electrons efficiently). Three types of such ion-molecule reactions which we have studied are charge transfer, e.g.,

$$O^+ + O_2 \rightarrow O + O_2^+ \tag{2.5}$$

and

$$He^+ + H_2 \rightarrow \begin{cases} He + H_2^+ \\ He + H^+ + H \end{cases} \tag{2.6}$$

atom transfer, e.g.,

$$O^+ + N_2 \rightarrow NO^+ + N \tag{2.7}$$

and association, e.g.,

$$H^+ + H_2 + H_2 \rightarrow H_3^+ + H_2 \tag{2.8}$$

The reactions involving O^+ are of importance in the earth's F_2 ionospheric region, while those involving H^+ and He^+ are important in the ionospheres of the outer planets such as Jupiter.

The method by which one studies these reactions in the laboratory involves use of a drift tube-mass spectrometer apparatus such as is shown schematically in Fig. 2.9. Using reaction (2.6) as an illustration, He^+ ions are generated by electron bombardment of helium (parent) gas in the ion source; they are then admitted to the drift region where a small amount of hydrogen (reactant) gas is mixed with an inert buffer gas (helium). The parent He^+ ions are made to drift across this region by a uniform electric field E, occasionally undergoing reactive collisions to form H^+ and H_2^+ product ions as they do. The parent and product ions reaching the far end of the drift region are analyzed by means of the quadrupole mass spectrometer and, from the disappearance of parent ions and formation of product ions, the reaction rate is determined.

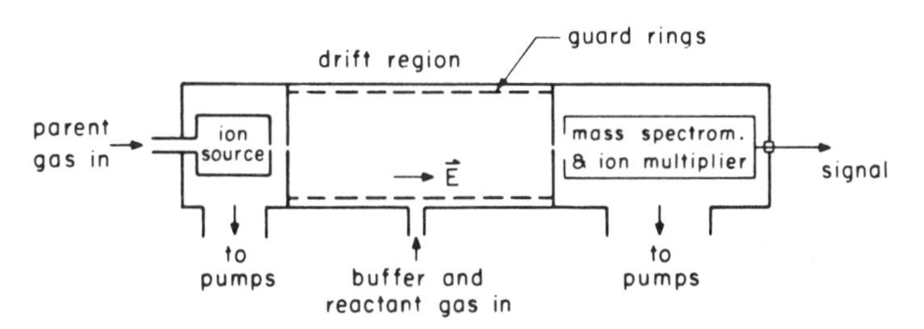

Fig. 2.9 Simplified diagram of a drift-tube mass-spectrometer apparatus used to determine ion-molecule reaction rates.

For the F_2 region of the earth, the conversion of O^+ by reactions (2.5) and (2.7) occurs at a rather slow rate (Johnsen and Biondi, 1973) (a point we shall return to later) and, in Jupiter's atmosphere, the conversion of H^+ to H_3^+ by reaction (2.8) occurs with appreciable probability only in the lower, denser regions, (Johnsen, Huang, and Biondi, 1976), while He^+ reacts very slowly (Johnsen and Biondi, 1974) with H_2 via reaction (2.6) and then forms H^+ rather than H_2^+. Studies of ion-molecule reactions of this type, together with the electron-ion recombination rate determinations discussed earlier, have formed the basis of the laboratory program to aid in modeling and understanding the behavior of the ionospheres of the earth and of the other planets.

4 FIELD OBSERVATIONS

With these laboratory programs, which continue actively to the present time, how were we lured from the comfort and familiarity of our laboratory surroundings into the sometimes difficult and inhospitable conditions involved in field observations? For me, our success in detecting the dissociative line broadening in recombination studies in neon was instrumental in leading to our venturing forth into the field.

Many years ago, a feeble red glow was detected in the night sky, a glow that was eventually identified as arising from the 6300A "forbidden" transition, $^1D \rightarrow {}^3P$, of atomic oxygen (see Fig. 2.10, lower right). Among the various hypotheses advanced concerning the origin of these excited oxygen atoms in the ionosphere, dissociative recombination between O_2^+ ions and electrons in the F-region was proposed, i.e.,

$$O_2^+ + e^- \rightleftharpoons (O_2^*)_r \rightarrow O^* + O^{*\prime} \quad (2.9)$$

where the product oxygen atoms can be formed in the various states ^3P, ^1D, as ^1S, as indicated in Fig. 2–10. Encouraged by our success in the

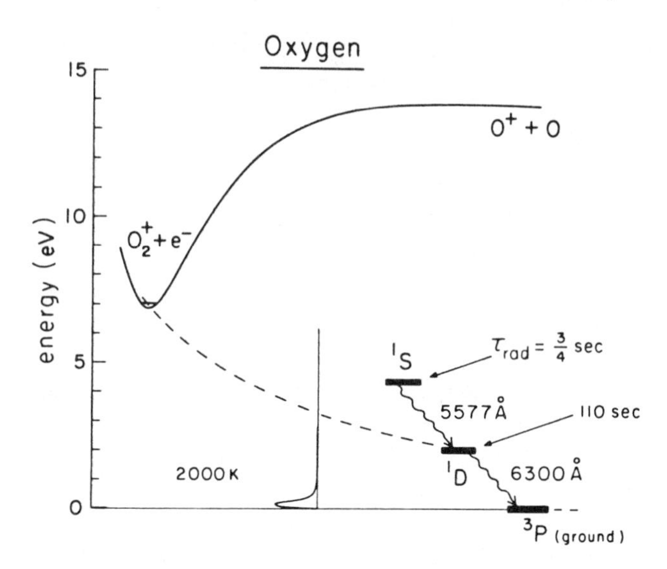

Fig. 2.10 Schematic representation of the energy levels of O_2^+ and of the low-lying states of atomic oxygen.

laboratory line-broadening studies, we embarked on similar studies of the 6300Å emission from the upper atmosphere to test whether or not dissociative recombination was indeed the origin of the oxygen red line.

There were, however, two major problems: (1) the radiation intensity from the night sky is far weaker than the feeble afterglow radiation in the laboratory studies, and (2) the fast excited $O^*(^1D)$ atoms have a radiative lifetime of 110 sec; thus there is a good chance that, before radiating, they slow down as a result of collisions with the atmospheric molecules, and so the dissociative broadening signature is lost. To overcome the first problem we designed a much larger aperature Fabry–Perot interferometer than used in the laboratory studies. As for the second, models of the nightglow radiation indicated that occasionally the emission might come from a high enough altitude that the excited oxygen atoms, moving in a near-vacuum, would not be slowed by collisions before radiating.

With a National Science Foundation grant we purchased a small van in which to mount our electronic equipment and constructed an 80 mm aperature interferometer housed in a telescope-like mount. Our first observations of the nighttime ionosphere were made from a nearby mountaintop in Pennsylvania in 1964. These first trials were remarkably unsuccessful—we were unable to detect the atomic oxygen line profile in the nightglow—but we learned a great deal about improving our interferometer's stability and detection sensitivity.

The following year a National Science Foundation grant to the University of Pittsburgh's aeronomy group enabled us to have built a permanent Airglow Observatory about 60 miles from the University atop Laurel Ridge, Pennsylvania. Here, in 1966, using a new 100 mm aperture interferometer we were successful in detecting and measuring the oxygen 6300Å nightglow line profile (Biondi and Feibelman, 1968), as indicated in Fig. 2.11. The results of several years' effort to detect dissociative broadening were negative—the nightglow profile exhibits a purely thermal doppler shape. However, this result should not be interpreted as eliminating dissociative recombination as the source of the 6300Å nightglow, since we now know that, at the altitudes where the bulk of the radiation originates, slowing of the excited atoms by collisions proceeds at a much greater rate than spontaneous radiation.

The adage that every cloud has a silver lining applies, since these thermal doppler profiles provided us with a remote thermometer for determining the temperature of the exosphere, a fundamental piece of ionospheric information that heretofore could only be inferred indirectly from a complicated analysis of satellite orbit decays. We now can measure accurately the change of exospheric temperature with time from sunset to sunrise and so determine the heating and cooling of the upper atmosphere in response to absorption of solar ultraviolet radiation. An added bonus obtained from the 6300Å line profile measurements is shown in Fig. 2.12, which illustrates our discovery of a sudden shift in the position of the line-center frequency during westward-looking observations at a low (30°) elevation angle (Biondi and Feibelman, 1968). This blue-shifted line was

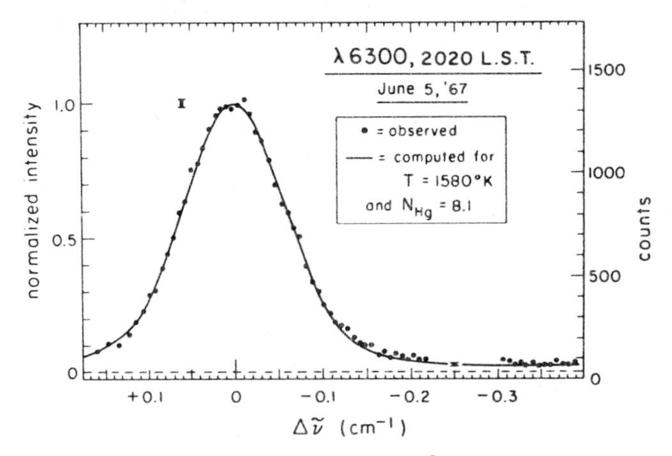

Fig. 2.11 Observed profile of the ionospheric 6300 Å nightglow line showing a purely thermal doppler broadening.

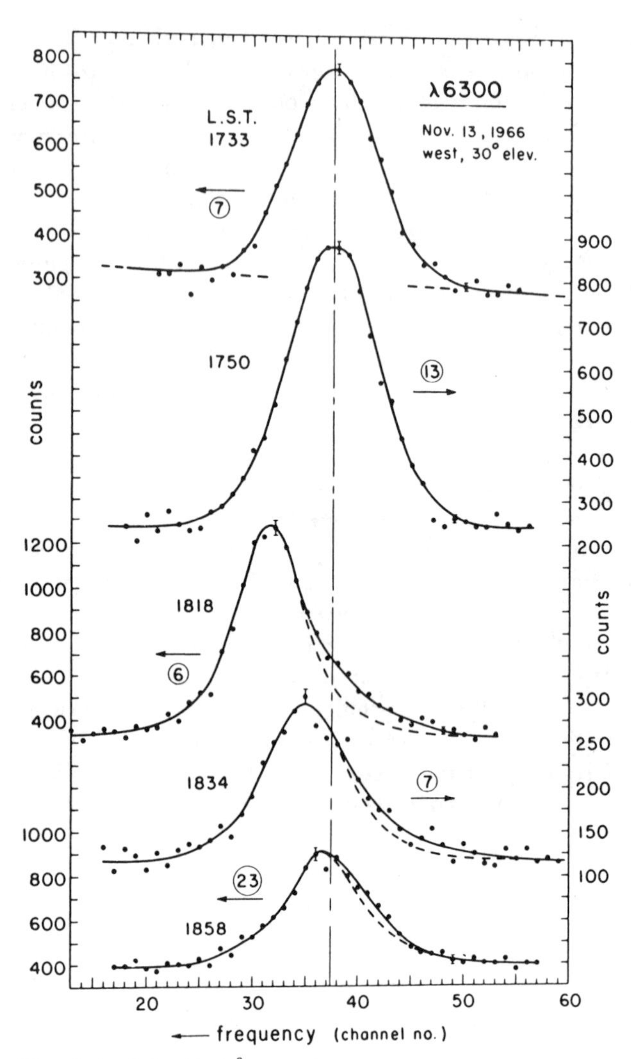

Fig. 2.12 Blue-shift of the 6300 Å nightglow line, indicating a substantial movement of the upper atmosphere (neutral wind) toward the observer.

interpreted as resulting from a very large ($\gtrsim 500$ m/sec) eastward wind during an ionospheric disturbance. Thus the interferometer measurements also provided us with an opportunity to determine the circulation patterns of the upper atmosphere, but we were unable to pursue this goal immediately because of a diversion of our efforts.

5 ARTIFICIAL PLASMA CLOUDS

So far, we have discussed the natural ionosphere, a largely invisible plasma that has profound effects on the propagation of the whole spectrum of radio waves. In order to study radio propagation through unstable plasmas, the government's Advanced Research Projects Agency (ARPA) wanted to create artificial (and visible) plasma clouds in the ionosphere by rocket releases of barium vapor at 150 to 200 km altitude. If one vaporizes 30 to 50 pounds of barium at this altitude, it quickly expands to a cloud several miles in diameter. If the release is just after twilight (so that from the ground the sky appears dark) the Ba cloud is made visible by its resonant scattering (fluorescence) of sunlight, as indicated schematically in Fig. 2.13. As suggested by the reactions in the figure, out of the continuous solar spectrum, neutral Ba atoms selectively absorb green photons to become excited to a resonance level and then reradiate these green photons in all directions, making the neutral cloud visible. Other solar photons have enough energy to ionize the Ba (either in a single-step process, as shown, or in two steps). The Ba^+ ions selectively absorb violet photons and reradiate them, making the plasma cloud visible and distinguishable from the neutral cloud.

In the auroral regions (high-latitudes) such an electron–Ba^+ ion plasma is driven to instability by the combination of the near-vertical magnetic field **B** of the earth and a horizontal electric field **E** present at times in the auroral zone. The **E** × **B** instability, as it is known, causes the plasma to bunch up into filamentary striations aligned along the earth's **B** field, and these plasma striations play havoc with radio wave propagation.

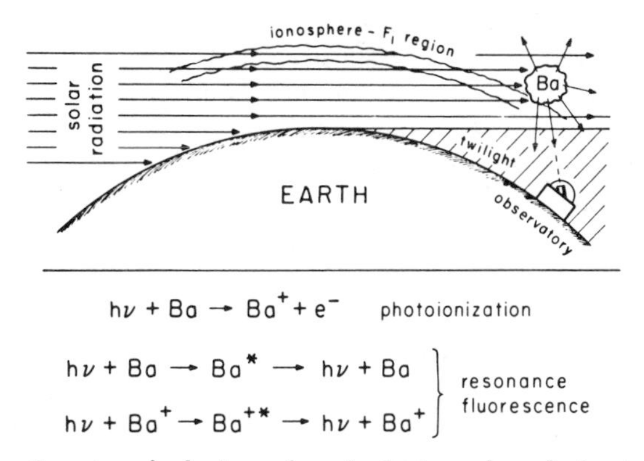

$$h\nu + Ba \rightarrow Ba^+ + e^- \quad \text{photoionization}$$

$$\left.\begin{array}{l} h\nu + Ba \rightarrow Ba^* \rightarrow h\nu + Ba \\ h\nu + Ba^+ \rightarrow Ba^{+*} \rightarrow h\nu + Ba^+ \end{array}\right\} \text{resonance fluorescence}$$

Fig. 2.13 Top: Geometry of a barium release in the ionosphere during twilight. Bottom: Reactions indicating photoionization and resonance fluorescence produced by solar photons $h\nu$.

To test theories of the onset and growth time of these instabilities one needs measurements of the plasma density at various points in the plasma. Were it not for the fact that the scattered solar photons are absorbed and reemitted many times in escaping from the Ba^+ cloud, a simple measurement of the total intensity of the scattered radiation would provide the desired Ba^+ "column" densities. With an interferometer it is possible to overcome this problem by tuning slightly off the center frequency of the optical transition, so that the absorption is small and the measured intensity is linearly proportional to the column density. ARPA, upon learning of our successful interferometric studies at Laurel Ridge, asked us to participate in their forthcoming barium release experiments in Alaska.

We modified one of our interferometers to include an optical tracking head and shipped everything to Fairbanks, Alaska in March, 1969. The apparatus was set up in the University of Alaska's auroral observatory atop Ester Dome, near Fairbanks.

The next figures are pictures taken with a bore-sighted camera that recorded the point in the neutral- or ion-cloud that we were studying. In Fig. 2.14, we see the spherical neutral cloud six minutes after the release of about 50 kg of barium at an altitude of about 180 km. The cloud has expanded to a diameter of about 8 km and ionization is taking place within, although the violet radiation marking the Ba^+ ions is largely obscured to the eye by the strong green radiation resonantly scattered by the neutral Ba atoms. (A trace of the ion cloud is visible as the somewhat irregular shape on the left edge of the spherical cloud.) Fortunately, the neutral- and plasma-clouds separate, since the neutral Ba atoms move in response to the motion of the neutral atmosphere, while the charged electrons and ions also respond to the $\mathbf{E} \times \mathbf{B}$ forces. Thus, at a later time (Fig. 2.15) one can see the separate plasma cloud on the left, that by now has been driven to instability and formed the filamentary striations of high plasma density aligned with the earth's magnetic field. With the Fabry–Perot interferometer we were able to measure the details of the Ba and Ba^+ spectral line shapes and so obtain the desired column densities at various points in the neutral and plasma clouds as the plasma instabilities developed (Hake, 1970).

While we were still in Alaska studying these artifical plasma clouds, we were approached by ARPA concerning a new experiment they were planning. For decades scientists had dreamed of producing controlled changes in the natural ionosphere of the earth in order to learn more about the detailed processes controlling the ionosphere. With its large resources ARPA was able to fund an experiment in which a powerful beam of radio frequency waves would be projected upwards into the ionosphere. The

Fig. 2.14 Neutral barium cloud at 180 km above the earth 6 min after the release. The cloud diameter is approximately 8 km.

radio frequency would be chosen to match the plasma frequency (refer to Fig. 2.2) in the F-region, so that the wave would be both reflected and rather strongly absorbed at that point. Theoretical calculations indicated that the megawatt of power available would cause the electron temperature T_e in the affected region to increase by about 30%.

We were asked whether we could detect this rise in T_e by some effect on F-region optical emissions. On the basis of laboratory studies of the formation of O_2^+ from O^+ charge transfer (Johnsen and Biondi, 1973) [reaction (2.5)] and dissociative recombination (Chanin, Phelps, and Biondi, 1962) of O_2^+ with e^- to form $O^*(^1D)$ [reaction (2.9)], we predicted that there would be an observable effect, as follows: In the F-region the 6300 Å line emission is thought to result from these two processes acting in sequence. The slow charge transfer reaction to convert the predominant O^+ ions to O_2^+ does not depend on T_e and therefore would be unaffected when the heating power was turned on. However, the recombination coefficient $\alpha(O_2^+)$, which varies approximately as $T_e^{-0.6}$, would decrease as the electron temperature rose, leading to an initial reduction in the $O^*(^1D)$ production rate and therefore in the 6300 Å intensity. Since the slow charge transfer process is rate limiting, the O_2^+ concentration would then build up and the 6300 Å intensity would

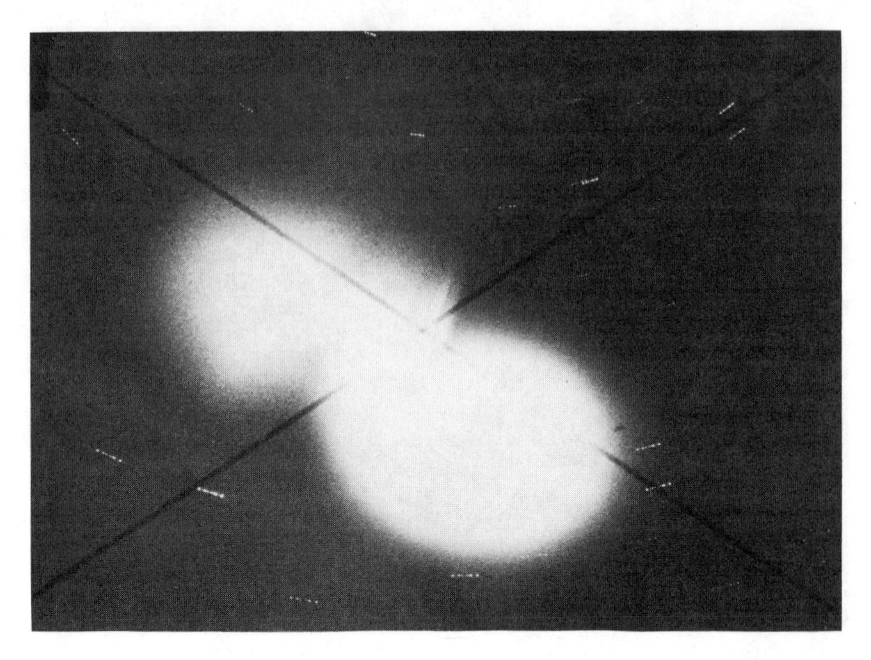

Fig. 2.15 Late time behavior. The elliptical neutral barium cloud (lower right) is now about 10 km in diameter. The striated ion cloud (upper left) has separated from the neutral cloud in response to $\mathbf{E} \times \mathbf{B}$ forces.

begin to return to its unperturbed values. When the heating power was turned off, the situation would reverse, with an initial increase in 6300Å intensity followed by a recovery toward the unperturbed value. Since this characteristic "suppression" signature would indicate the presence of ionospheric modification by the heating transmitter and would permit an estimate of the electron temperature-rise produced, we received funding to construct a sky-mapping photometer to detect the regions where the nightglow radiation was affected.

Our initial measurements were carried out near Denver, Colorado in conjunction with the powerful transmitters of the Institute for Telecommunication, and we successfully detected the predicted 6300Å suppression signatures (Biondi, Sipler, and Hake, 1970). However, there was a lack of sufficient instrumentation to determine many of the relevant ionospheric parameters for a detailed comparison with the ionospheric model. Thus, under National Science Foundation sponsorship, a similar experiment was later set up at Arecibo, Puerto Rico, where the world's largest radio telescope "mirror," some 1000 feet in diameter, is located. When used as a backscatter radar antenna this installation provides detailed information on the variation with altitude of the electron density, electron temperature, plasma waves (instabilities), etc., in the ionosphere.

Our photometer was housed atop a hill adjacent to the radio telescope in Arecibo's airglow observatory building. Here, in a program which continues to the present, we have been studying the response of the F-region to ionospheric modification by less-powerful (about 100 kW), but more concentrated beams of radio energy.

A recent example of 6300Å supression transients observed at Arecibo (Sipler and Biondi, 1978) is given in Fig. 2.16. As predicted, the intensity decreases when the heating power is turned on and increases when it is turned off. The solid lines, that reproduce the observed variations, are the model predictions using backscatter radar values of n_e and T_e and the laboratory values of $\alpha(O_2^+)$ and its dependence on T_e. Thus, the objective that prompted our initial venture into field observations—to show that the 6300Å line emitted from the F-region originates from dissociative

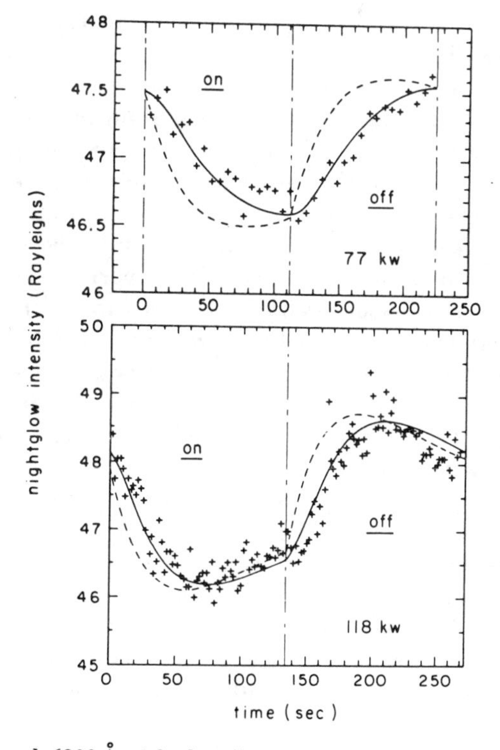

Fig. 2.16 Observed 6300 Å nightglow "suppression" transients resulting from ionospheric electron heating by ground-based radio transmitters at Arecibo, Puerto Rico. The solid lines are the theoretically predicted behavior.

recombination of O_2^+ ions—was finally achieved. These ionospheric modification studies not only permit us to learn more about the various atomic collision processes occurring in the F-region, but also have enabled us to detect effects of plasma instabilities generated by the heating radio wave (Sipler and Biondi, 1972).

Most recently, we have been pursuing the study of upper atmosphere circulation by using the doppler shifts in the 6300Å line positions (such as shown in Fig. 2.12) to determine the neutral wind vectors in the exosphere some 300 km above the earth's surface. The principal forces that drive the neutral atmosphere arise from the regular heating cycles caused by absorption of solar ultraviolet radiation and from occasional "storm" periods when large numbers of energetic particles precipitate into the atmosphere. From our observatory on Laurel Ridge, Pennsylvania, we have been engaged in cooperative neutral wind studies with Fritz Peak, Colorado to determine the changing atmospheric circulation patterns at midlatitudes. The measurements are then compared with hydrodynamic model predictions of the atmospheric flow in response to pressure gradients produced by the solar heating.

The ability to measure exospheric wind vectors led to our participation in a concerted study of the equatorial ionosphere during August, 1977. The equatorial ionosphere is believed to exhibit yet another form of plasma instability caused by the vertical gravitational forces acting together with the equatorial magnetic field **B** (which is horizontal), the so-called $\mathbf{g} \times \mathbf{B}$ instability. As with the auroral $\mathbf{E} \times \mathbf{B}$ instability in the barium plasmas, the equatorial $\mathbf{g} \times \mathbf{B}$ instability causes large portions of the F-region plasma to form striated regions (within "plasma bubbles") which seriously affect communication between satellites and earth-based stations.

According to theory the vertical and horizontal movement of these plasma bubble regions is controlled by the neutral wind impinging on the plasma; thus, our Fabry–Perot doppler shift measurements could determine one of the important ionospheric parameters. For our part, we were anxious to measure equatorial exospheric winds, since a rather unusual phenomenon named "superrotation" had been postulated to explain observed changes in satellite orbit inclinations. This superrotation, that was supposed to involve a constant eastward motion of the equatorial upper atmosphere (at a speed of several hundred meters/sec at 300 km altitude), was very difficult to understand, and some rather extreme hypotheses such as incoming meteoroids had been advanced to explain its origins.

In August 1977 we shipped an improved version of one of our interferometers to Roi-Namur Island in the Kwajelein Atoll, that lies near the equator several thousand miles west of Hawaii across the International

Date Line. Here we were successful in measuring the F-region neutral winds from the doppler shifts in the 6300Å nightglow line. Examples of the observations are given in Fig. 2.17 for the nights of August 20 and 21, 1977 (12 UT = local midnight). The wind vector is found to be of modest magnitude (50 to 150 m/sec) and is observed to turn southward and diminish in the late night, thus dispelling the hard-to-understand notion of a constant, eastward superrotation velocity. The observed behavior resembles qualitatively the circulation patterns calculated from the hydrodynamic pressure gradient models; thus our data provide a better source of atmospheric pressure gradient information than has heretofore been available.

To summarize, these field studies of the earth's upper atmosphere, which advance the understanding of our environment, owe a large debt to the laboratory experiments for providing much of the atomic collision data needed to unravel the complex phenomena which are observed. There is a continuing interplay between atmospheric observations and laboratory studies, each benefitting from the insights provided by the other.

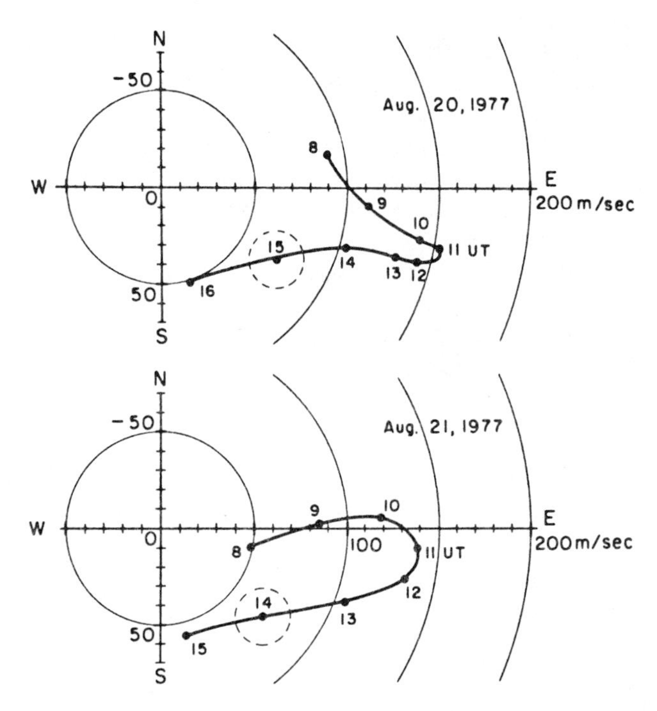

Fig. 2.17 Measured positions of the tip of the equatorial F-region neutral wind vector as a function of time during two nights in August 1977 (12 UT = local midnight).

REFERENCES

Bates, D. R., 1950a, *Phys. Rev.*, **77**, 718.

Bates, D. R., 1950b, *Phys. Rev.*, **78**, 492.

Bates, D. R. and H. S. W. Massey, 1946, *Proc. Roy Soc. (London)*, **A187**, 261.

Bates, D. R. and H. S. W. Massey, 1947, *Proc. Roy. Soc. (London)*, **A192**, 1.

Biondi, M. A., 1956, *Rev. Sci. Instr.*, **27**, 36.

Biondi, M. A. and S. C. Brown, 1949a, *Phys. Rev.*, **75**, 1700.

Biondi, M. A. and S. C. Brown, 1949b, *Phys. Rev.*, **76**, 1697.

Biondi, M. A. and W. A. Feibelman, 1968, *Planet. Space Sci.*, **16**, 431.

Biondi, M. A., D. P. Sipler, and R. D. Hake, Jr., 1970, *J. Geophys. Res.*, **75**, 6421.

Chanin, L. M., A. V. Phelps, and M. A. Biondi, 1962, *Phys. Rev.*, **128**, 219.

Connor, T. R. and M. A. Biondi, 1965, *Phys. Rev.*, **140**, A778.

Frommhold, L. and M. A. Biondi, 1969, *Phys. Rev.*, **185**, 244.

Hake, R. D. Jr., 1970, "Interferometric observations of OI, BaI and BaII optical emissions from the upper atmosphere," Ph.D. thesis, University of Pittsburgh.

Huang, C. M., M. A. Biondi, and R. Johnsen, 1976, *Phys. Rev.*, **A14**, 984.

Johnsen, R. and M. A. Biondi, 1973, *J. Chem. Phys.*, **59**, 3504.

Johnsen, R. and M. A. Biondi, 1974, *J. Chem. Phys.*, **61**, 2112.

Johnsen, R., C. M. Huang, and M. A. Biondi, 1976, *J. Chem. Phys.*, **65**, 1539.

Kasner, W. H. and M. A. Biondi, 1968, *Phys. Rev.*, **174**, 139.

Leu, M. T., M. A. Biondi, and R. Johnsen, 1973a, *Phys. Rev.*, **A8**, 413.

Leu, M. T., M. A. Biondi, and R. Johnsen, 1973b, *Phys. Rev.*, **A8**, 420.

Mehr, F. J. and M. A. Biondi, 1969, *Phys. Rev.*, 181, 264.

Rogers, W. A. and M. A. Biondi, 1964, *Phys. Rev.*, **134**, A1215.

Sipler, D. P. and M. A. Biondi, 1972, *J. Geophys. Res.*, **77**, 6202.

Sipler, D. P. and M. A. Biondi, 1978a, *Geophys. Res. Lett.*, **5**, 373.

Sipler, D. P. and M. A. Biondi, 1978b, *J. Geophys. Res.*, **83**, 1519.

Weller, C. S. and M. A. Biondi, *Phys. Rev. Lett.*, **19**, 59.

CHAPTER 3

Applications and Needs

ARTHUR V. PHELPS
Joint Institute for Laboratory Astrophysics
National Bureau of Standards and University of Colorado
Boulder, Colorado

1 INTRODUCTION

This is a review of the state of knowledge of the electron-molecule scattering with particular attention to the applications of that knowledge to technology. We also consider the needs for further knowledge of electron-molecule scattering in order to meet current and foreseeable technology. For convenience, the review is divided into three main parts. The first is a discussion of applications and the state of knowledge involving air and its constituent gases. Note that we do not discuss the earth's atmosphere or ionosphere, etc., since these are thoroughly covered in Professor Biondi's paper. The second part is concerned with the role of electron-molecule scattering in the area of energy generation, e.g., MHD generators and isotope separation. Finally, we discuss some of the applications and needs in the area of gaseous lasers utilizing electron excitation of molecules.

2 AIR

We begin this discussion with a review of the electron-molecule collision in air with particular attention to the questions of application, the state of our present knowledge, and the need for additional data. Without discussing all of the possible applications and information about electron collisions with molecules of air, we turn our attention to chemical production, electrical breakdown, air quality modification and monitoring.

2.1 Chemical Production

A schematic diagram of a Siemens ozonizer (McTaggart, 1967; Bell, 1971) is illustrated in Fig. 3.1 with some of the principal reactions that take place in the commercial production of ozone. The capacitor plates are driven by a high voltage ac generator. A dielectric material covers the electrodes and a series of short electrical discharges is produced in the space between the electrodes. The principal initial reaction is the dissociation of the oxygen molecule. The resulting atoms associate with neutral diatomic oxygen molecules to produce the ozone. Ozone can be destroyed in neutral reactions and the negative ion formation and further dissociation in the electron collision reactions are of interest. Unfortunately there is currently a very large factor of 3 to 10 uncertainty in the rate coefficients for the dissociation of O_2 by electrons.

A second application is the anodization of metals to produce oxygen films (Chang and Sinha, 1976). According to one model the first step is to

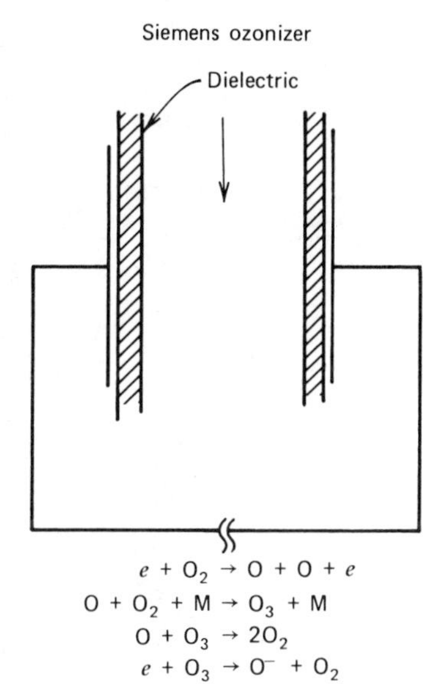

Fig. 3.1 Schematic diagram of a Siemens ozonizer.

produce a dissociative attachment by electron impact

$$e + O_2 \rightarrow O + O^-$$

The resulting negative ions are pulled toward the anode by an electric field where they react with the metal surface:

$$O^- + M \xrightarrow{S} MO + e(\text{conduction})$$

This is a very simplified model, but it illustrates a type of chemical reaction that can occur on surfaces. There are many other examples such as depositing amorphous silica layers; some suggestions have been made that this can be done using silane in an electrical discharge.

2.2 Electrical Breakdown

Studies have been made on the role of electron-molecule collisions in high voltage transmission systems. As an illustration, Fig. 3.2 shows some very recent data of the conditions for propagation of streamers formed in an atmospheric pressure discharge [data from Phelps and Griffiths (1976)]. A streamer occurs when electrons are accelerated in a high enough

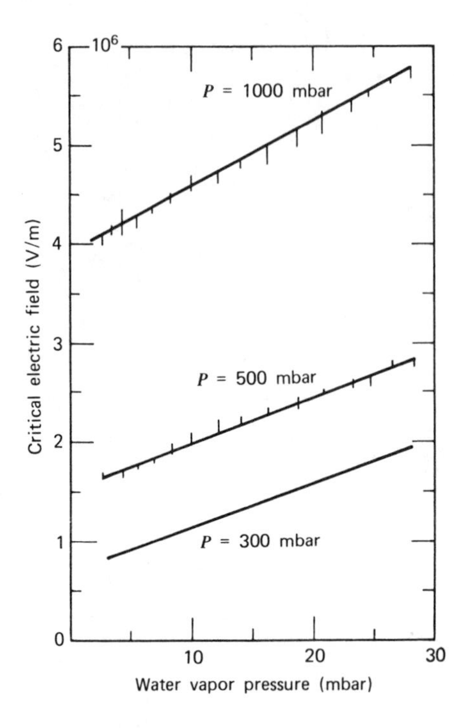

Fig. 3.2 Streamer formation in air and water vapor.

electric field so as to produce sufficient ionization to highly perturb the initial electric field. In the process of acceleration the electrons produce photons which move away from the discharge and produce new electrons. These in turn are accelerated and, if the field is sufficient, a self-propagating discharge results. The data in Fig. 3.2 for air and water vapor show the field required for streamer propagation at various pressures of water vapor. The effect is rather large and Phelps proposes that the effect is due to the role of water vapor in stabilizing electron attachment in oxygen.

2.3 Air Quality Modification and Monitoring

One of the more common applications in the area of air quality modification is in air cleaning using electrostatic precipitators. The technique is very similar to ozone production. The geometry is somewhat different, but basically a small corona discharge that produces the charged particles, also produces ozone. In fact, these devices when used in offices and homes have to be run below maximum efficiency to keep down the ozone production.

An example of air quality monitoring is the use of electron beam

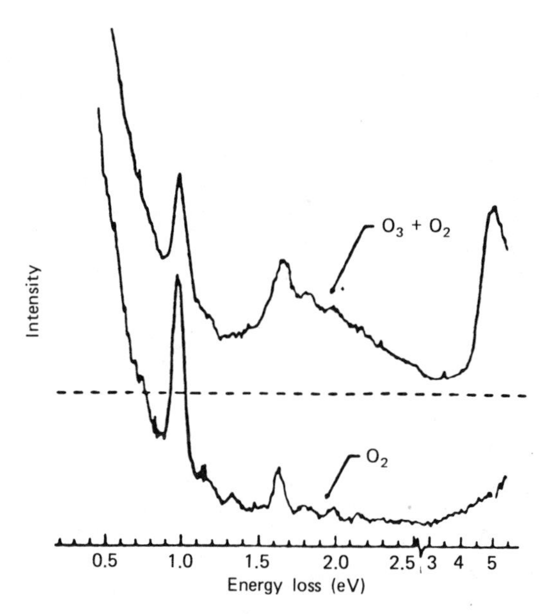

Fig. 3.3 Energy-loss spectrum of ozone-oxygen mixtures and pure oxygen.

scattering to measure the concentrations of small amounts of added gases. Typical of this approach is the work at the National Bureau of Standards (Swanson and Celotta, 1975) illustrated in Fig. 3.3. The top curve is the electron energy-loss spectrum of an ozone-oxygen mixture. The bottom curve is the energy-loss spectrum of pure oxygen. The signal due to the ozone is clearly visible. This type of scattering experiment can thus be used to obtain high sensitivity for the detection of impurities in air.

2.4 Present State of Knowledge

What is the state of our knowledge regarding electron collisions with molecules of air, nitrogen, oxygen and water vapor? Most important was the work done by Schulz on the vibrational excitation of the nitrogen molecule, the results of which are shown in Fig. 3.4 (Schulz, 1964). These data show that the cross section for vibrational excitation for various states of the nitrogen molecule is very large and dominates the electron energy behavior in the volt or several volt region, for example at electrical breakdown. At higher energies, electronic excitation becomes important. Early work by Rees and Jones (1973) on the excitation and ionization of nitrogen in the 1 to 1000 V region is summarized in Fig. 3.5. Much more recent experimental work at the Jet Propulsion Laboratory at the Califor-

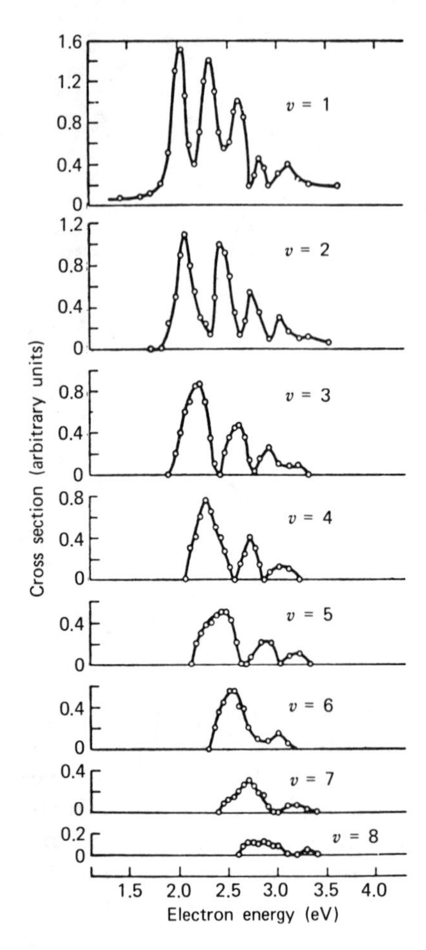

Fig. 3.4 Vibrational excitation of nitrogen.

nia Institute of Technology and theoretical work at Los Alamos has filled in much more detail of electron impact excitation of the electronic states of nitrogen (Cartwright, Trajmar, Chutjian, and Williams, 1977; Chutjian, Cartwright, and Trajmar, 1977). These studies have been very useful in understanding such things as the nitrogen laser.

Let us now turn to the oxygen molecule. Electron attachment and vibrational excitation are shown in Fig. 3.6. This figure is a composite of the experimental results of Spence and Schulz (1972) for the three-body attachment process and the vibrational excitation cross section from the work of Linder and Schmidt (1971). It can be seen that even at low energies there is a great deal of structure. Further work has shown that the width of peaks is limited by the apparatus and that the widths of some of

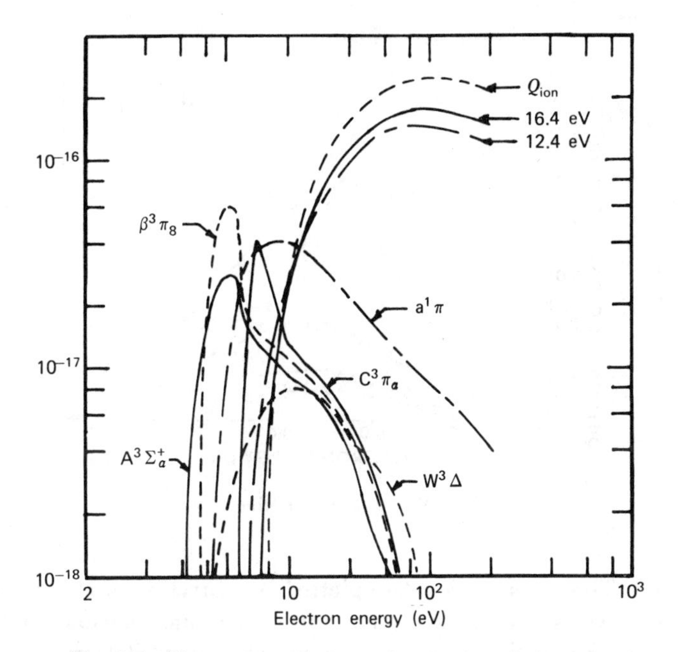

Fig. 3.5 Electronic excitation and ionization of nitrogen.

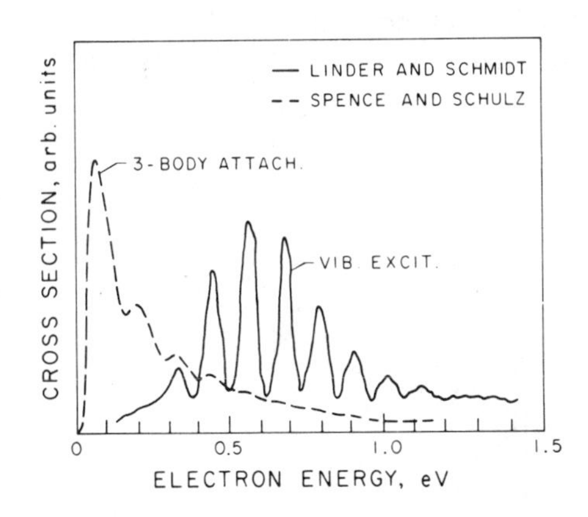

Fig. 3.6 Electronic attachment and vibrational excitation in oxygen.

these resonances are in the micro-volt region, corresponding to lifetimes of 10^{-10} sec.

In the higher energy region, the work of Wong, Boness, and Schulz (1973) is shown in Fig. 3.7. Here the vibrational excitation of the oxygen molecule shows as a very broad resonance around 9 eV. In an attempt to form a complete picture of electron motion and inelastic collisions in the oxygen molecule, a summary has been prepared (Lawton and Phelps, 1977)

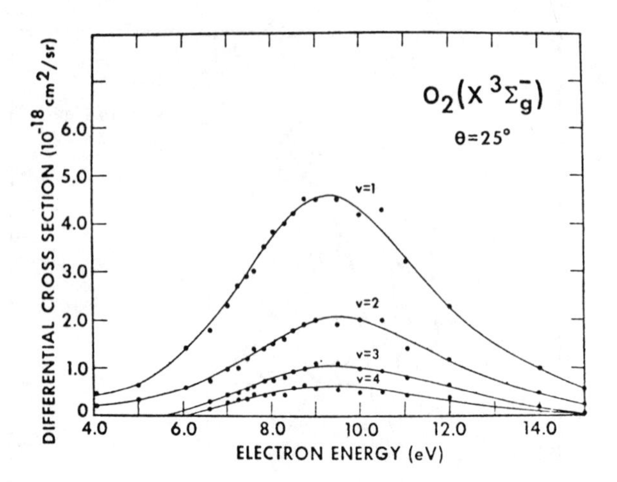

Fig. 3.7 Vibrational excitation of oxygen.

in Fig. 3.8. The cross sections are plotted as a function of electron energy. The top curve is the cross section for momentum transfer, and at the bottom are shown the peaks due to vibrational excitation, the various metastable and electronic excitation processes and finally, ionization at higher energies. It would appear that by looking at electron beam data we have been able to come up with a moderately complete description of electron behavior in oxygen.

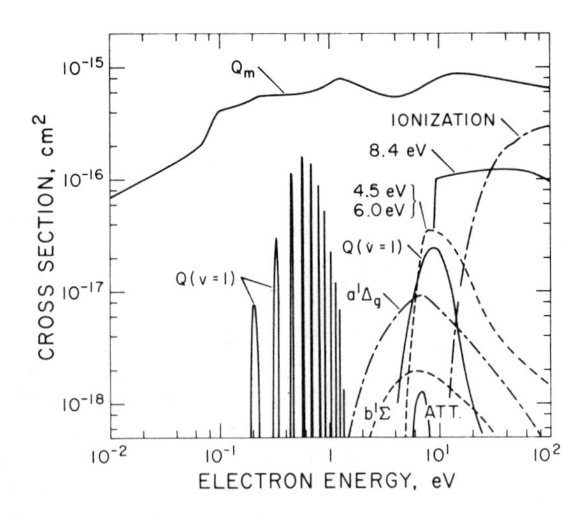

Fig. 3.8 Cross sections for electron excitation of oxygen.

We now compare calculated electron transport coefficients with the measurements in Fig. 3.9, where the electron collision frequency is shown as a function of the characteristic energy (the electron temperature if the electrons had a Maxwellian velocity distribution). The experimental data are represented by the solid curve and the points are calculations based on cross sections such as those shown in Fig. 3.8. This figure clearly illustrates some disagreement. If the vibrational excitation cross sections given by Linder and Schmidt (1971) are used, the calculated values are low compared with the experimental data in the 0.1 to 1 V energy range. On the other hand, if we leave Linder and Schmidt's cross sections as measured, but use the predictions of Geltman and Takayanagi (1966) for rotational excitation in oxygen, the high triangular points are obtained. It is obvious that we do not have a complete description of electron motion in oxygen in the few volt energy range. In terms of the practical application of these low energy cross sections in oxygen it is well to point out that oxygen dominates the behavior of low energy electrons in air.

Another example from oxygen is the recent work by Lawton and Phelps (1977) on the excitation of the $b^1\Sigma_g^+$ state of this molecule. This is shown in Fig. 3.10. This excitation coefficient is the number of excitation events per centimeter of drift in the field direction, similar to the Townsend ionization coefficient. If this coefficient is multiplied by the electron drift velocity, one gets the usual first-order rate coefficient used by chemists.

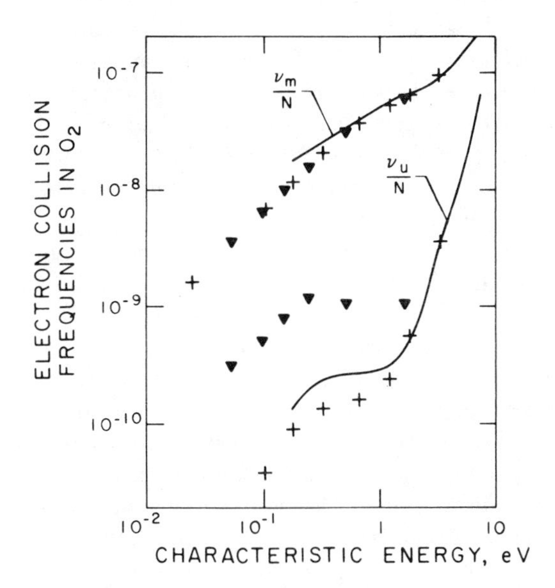

Fig. 3.9 Electron collision frequency as a function of characteristic energy in O_2.

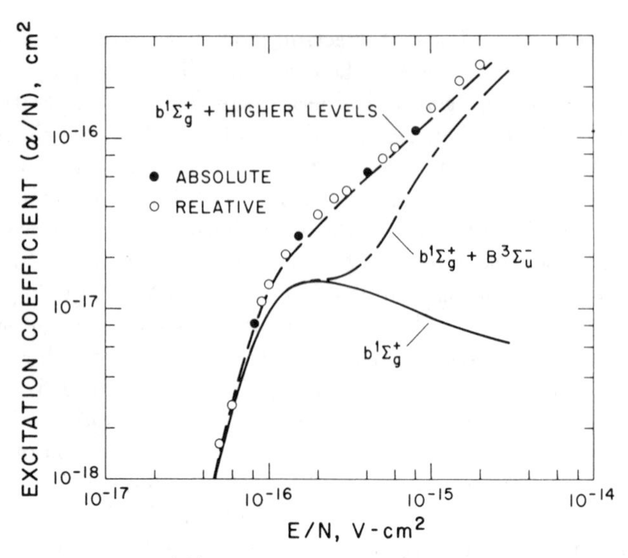

Fig. 3.10 O_2 $(b^1\Sigma_g^+)$ excitation by electrons as a function of E/N.

The experiment is indicated by the points and the curve represents various theoretical calculations based on the cross sections previously shown. The solid curve is calculated using the results of electron beam experiments by Linder and Schmidt (1971) and by Trajmar and co-workers (Trajmar, Cartwright, and Williams, 1971; Trajmar, Williams, and Kupperman, 1972; Hall and Trajmar, 1975). The fit at low energy is very good. At the higher energy range, the solid curve is well below the experiments. There is one additional process which leads to production of this state. It is the electron-impact dissociation of the molecule to produce singlet *D* atoms. The singlet *D* atoms collide with ground state molecules to produce the metastable. The efficiency values found in the literature vary markedly, but if we use 100% efficiency the long–short dashed curve is obtained. It does not fit badly at the higher mean energies, but in the 2 V range a factor of 3 discrepancy is evident. This discrepancy is not explained.

In Fig. 3.11 we draw together electron energy exchange frequencies for water vapor, dry air, oxygen, and nitrogen as a function of energy (Phelps, 1972). As we have mentioned before, the vibrational excitation of oxygen dominates in the few 0.1 V energy range. When water vapor is present, near sea level, the energy loss to water vapor dominates at energies below about 0.1 eV. Unfortunately the accuracy for water vapor is poor. The uncertainty is probably around a factor of 2 at the lower energies and perhaps worse at higher energies.

The state of our knowledge of attachment and ionization in air is shown

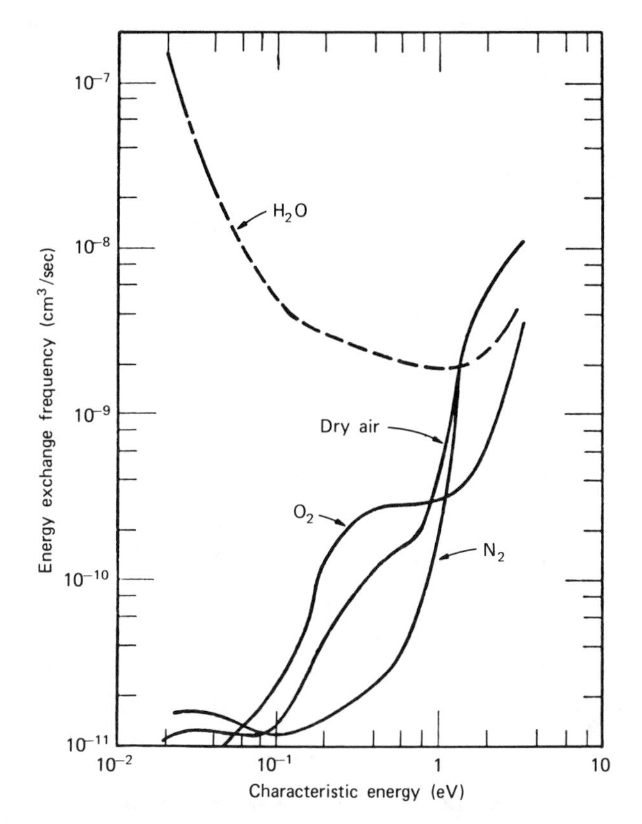

Fig. 3.11 Electron energy exchange frequencies as a function of characteristic energy for O_2, N_2, dry air, and H_2O.

in Fig. 3.12 plotted as a function of the electric field to gas density ratio. This covers an energy range from about 2 eV mean energy up to about 10 eV. As before, the coefficients are the number of events per centimeter of drift in the electric field direction. The solid curves are predictions and the points are experimental. Clearly the agreement is poor, possibly because associative detachment has been neglected. This would not be an electron process but one where an oxygen negative ion collides with a nitrogen molecule. This is a process that was studied in some detail by George Schulz and co-workers at Yale and at the NOAA laboratory in Boulder.

In addition to the needs for resolving the discrepancies cited in the preceding discussion, there are numerous other technological problems requiring additional information regarding electron collisions with molecules. Thus, the details of buildup of a discharge is an important problem,

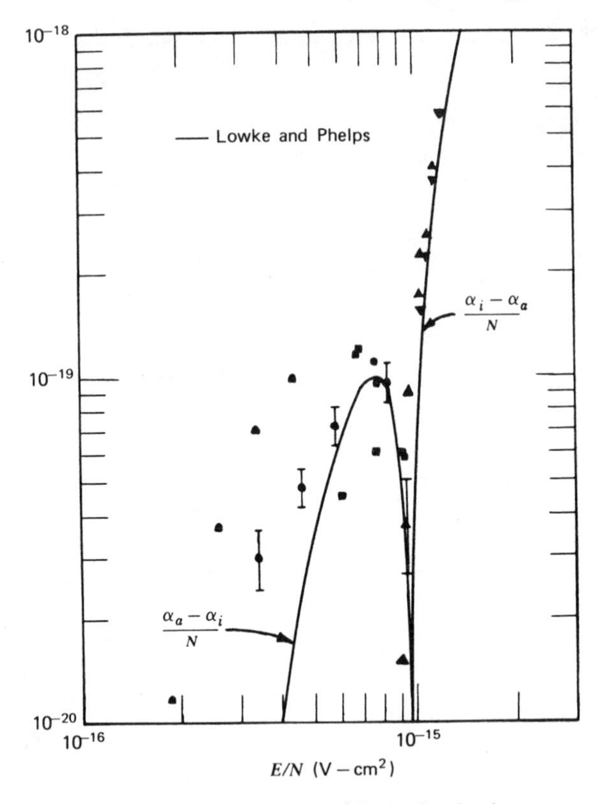

Fig. 3.12 Attachment and ionization in air.

for example, in long spark breakdown in high voltage systems or in tanker explosions. A number of people have been concerned with the question: "How much energy is necessary to ignite an explosive mixture in air?" If the discharge has too small an energy input, ignition does not occur. Here again, additional information regarding the effects of electron collisions with vibrationally excited molecules is necessary. At present we do not have such information in adequate detail.

3 ELECTRON-MOLECULE COLLISIONS IN ENERGY GENERATION

3.1 Magnetohydrodynamic Generators

First let us discuss the generation of electric power by means of magnetohydrodynamic generators. Basically a MHD generator makes use of a hot ionized gas. The current interest is primarily in hot gases produced by burning coal to which the alkali metal potassium has been added to obtain ionization. These hot gases flow rapidly through a magnetic field that separates the positive and negative charges to produce an electric field. To know how much power can be extracted, the amount of current that flows as a result of this electric field has to be determined. This is a question of the resistivity of the medium. In Fig. 3.13 is shown the contribution of various gases to the resistivity in a gas that simulates that from burning coal seeded with a fraction of a percent of potassium (Spencer and Phelps, 1976). The fractional resistivity is plotted as a function of the temperature of the gas. At lower temperature (the ones usually obtained in present day systems), the dominant contribution to the resistivity is electron-molecule scattering by the major component of the input gas nitrogen and by water vapor produced in the burning process. Somewhat lower in magnitude is the contribution from carbon dioxide, carbon monoxide and potassium itself. It is conventional to assume that electron scattering by nitrogen,

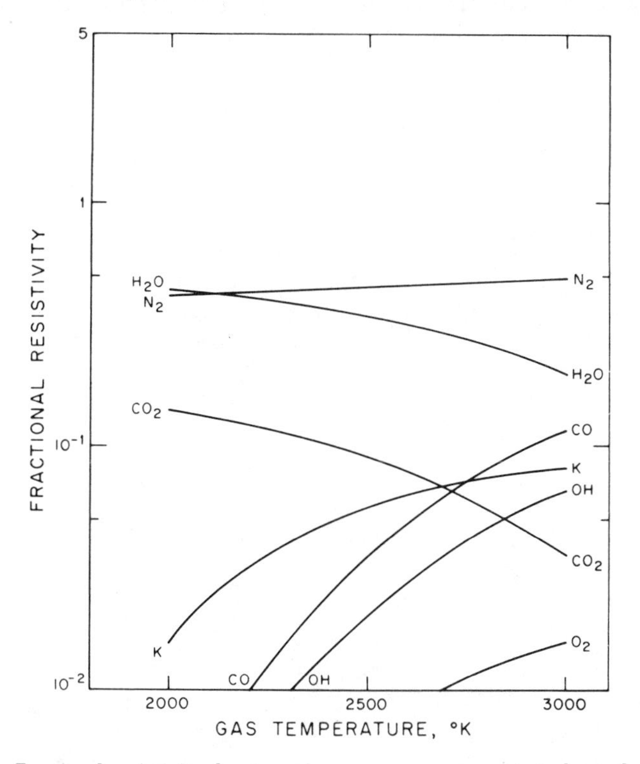

Fig. 3.13 Fractional resistivity due to various gaseous components in burned coal seeded with potassium.

carbon dioxide and carbon monoxide are known in this energy range to about 10%. But no one has measured the change in the momentum transfer scattering cross section when vibrationally excited species appropriate to these temperatures, so that the 10% accuracy may be illusionary. Furthermore, one species has been left off of Fig. 3.13. That species is KOH. The uncertainty caused by this species is highlighted in Fig. 3.14. Even though potassium is present only in very low levels, the KOH molecule has such a large dipole moment that the electron scattering from it could be the dominant contribution to the resistivity. The solid line through the middle represents the estimate based on the Born approximation using work originally done by Altschuler (1957). The upper boundary of the shaded region represents an empirical correction applied on the assumption that the theory was off by the same amount as is water vapor. The lower boundary represents some theoretical guesses as to how much the theory could be in error.

There has been considerable effort, both in the laboratory and theoretically to study the electron scattering by highly polar molecules. Very few people have tried to solve the problem of electron scattering by

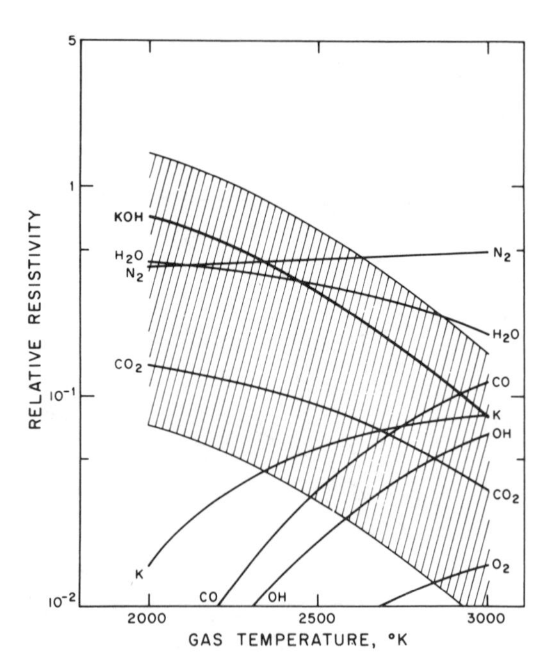

Fig. 3.14 Relative contributions of KOH to the resistivity of burned coal seeded with potassium.

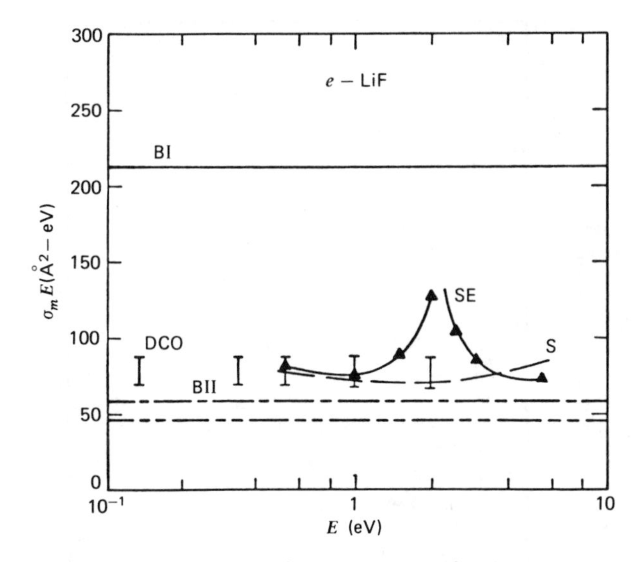

Fig. 3.15 Momentum transfer cross sections for electrons in LiF.

triatomic molecules. Most people simplify the KOH problem by studying molecules such as LiF. This molecule also has a very large dipole moment. Fig. 3.15 shows the momentum transfer cross section in this gas (Collins and Norcross, 1977). Actually what is plotted is the product of the cross section and energy as a function of electron energy. It is the lower energy part of these curves that are of interest in the burned coal study. Using the implications of these cross sections to the KOH seeded gas, the current estimation is that the actual effect of KOH on the resistivity puts its contribution somewhere in the middle of the shaded region of Fig. 3.14. The conclusion is that KOH is important although not dominant in determining the resistivity of a coal fired magnetohydrodynamic generator.

3.2 Isotope Separation for Fission Generators

Among the various methods of isotope separation, one is that using electron-molecule collisions. Typical of the work that led to this development is the work of Chantry (1969) shown in Fig. 3.16. Here is shown the relative cross section for electron collisions with N_2O resulting in the formation of O^- at temperatures between 295 and 1040°K. It shows a change in attachment rate or attachment cross section of many orders of magnitude as the internal energy increases.

Chen and Chantry (1978) have recently found similar phenomena in SF_6.

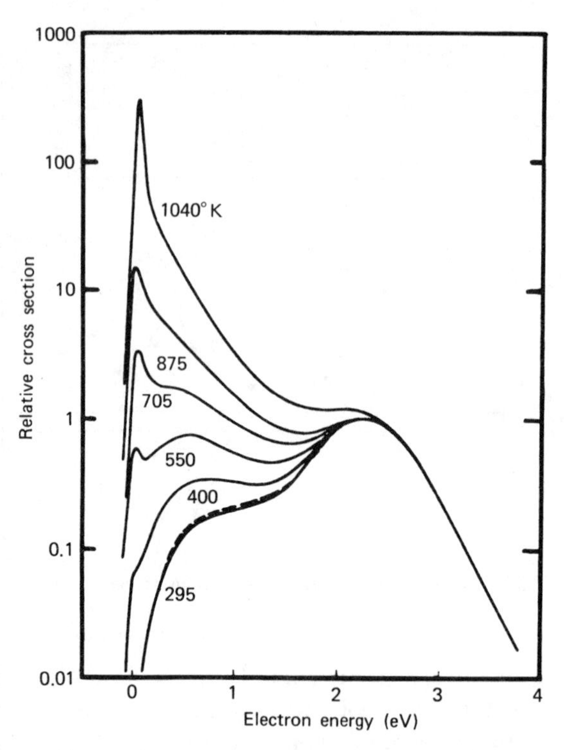

Fig. 3.16 O^- from N_2O at various temperatures.

Their work is shown in Fig. 3.17, where the current of SF_6 ions produced by electron impact is plotted as a function of electron energy. In general in SF_6 one sees a very low energy peak (called a zero energy peak) and then a shoulder up in the fraction of a volt range. Fig. 3.17 shows the change in the SF_5^- signals produced when S^{32} and S^{34} isotopes were used. Using specific lines from a CO_2 laser, they irradiated the SF_6 to produce vibrationally excited molecules which formed negative ions on electron collision. The top trace in Fig. 3.17 was with the laser present. A large peak is seen at zero electron energy and a small shoulder at the higher energies. With the laser turned off, the zero energy peak becomes much lower. The difference between these two curves gives the production of negative ions by the combined vibrational excitation using the laser and the electron impact attachment. The process is very specific to the particular isotope, the P28 line for S^{32} or P44 for S^{34}. This is a good illustration of possible isotope separation processes using electron collisions with molecules.

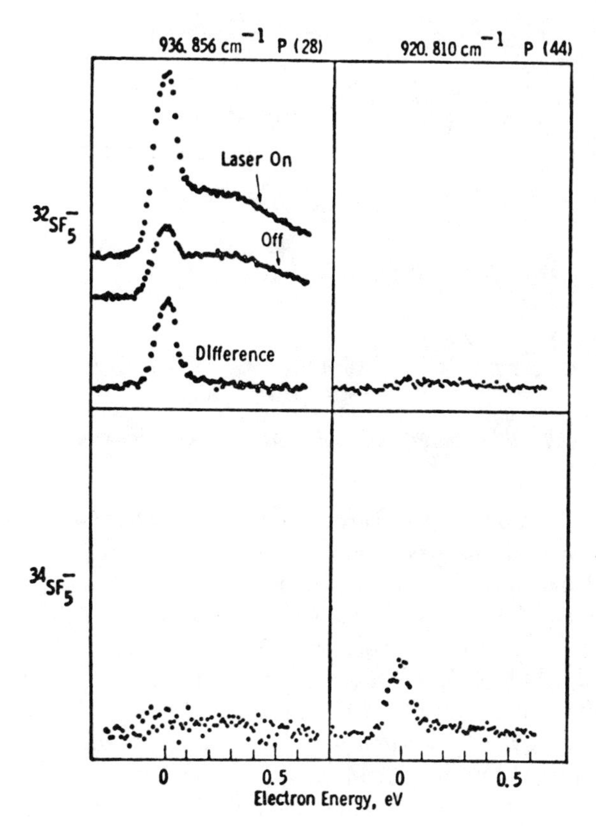

Fig. 3.17 Effect of CO_2 laser excitation of SF_6 on dissociative attachment in SF_6.

4 LASERS

4.1 Electron-Molecule Collisions in Lasers

First let us consider electron excitation of vibrational levels as in CO_2. There are several modes in which the CO_2 molecule can vibrate: the bending mode, the asymmetric stretch mode and the symmetric stretch mode. The vibrational excitation of these modes by electrons are shown in Fig. 3.18 from the work done in Schulz's laboratory (Boness and Schulz, 1974). The large resonances at 3.8 eV has been the subject of very thorough study. A simple Born approximation theory is quite adequate for the bending and symmetric stretch modes near threshold. On the other hand, the theory for the asymmetric stretch mode is in very poor shape.

With these types of information and knowledge of electronic scattering,

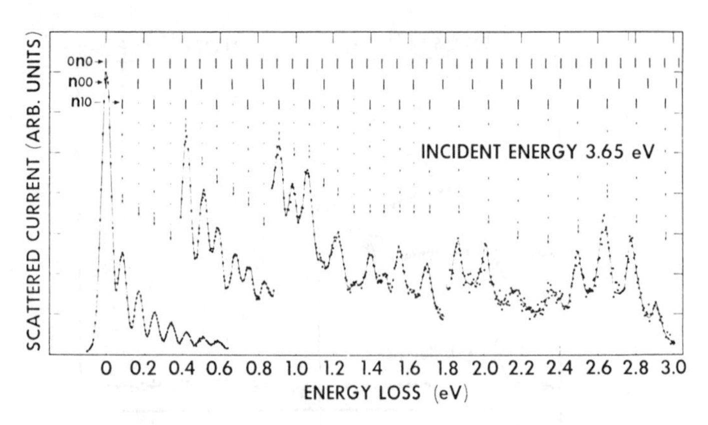

Fig. 3.18 Vibrational excitation by electron impact on CO_2 at 3.65 eV.

a moderately complete set of cross sections for electrons in CO_2 can be derived. It also allows predictions to be made of ionization coefficients. Figure 3.19 indicates the form of the ionization coefficient, again in terms of the number of events per centimeter of drift in the field direction as a function of the ratio of the density divided by the electric field (Phelps, unpublished). This kind of information can be used to make predictions regarding discharges in CO_2 lasers. In Fig. 3.20 the predictions of Denes and Lowke (1973) are compared with some of their experiments, with and without small amounts of water vapor. As can be seen, the electric field necessary to maintain the discharge is essentially independent of current. It results from a balance between attachment and ionization at the low current levels. At the higher current levels a new electron loss process sets in, dissociative recombination of electrons with the molecules, which in the presence of water probably means some kind of ion cluster.

The efficiency of excitation of the upper laser level for various CO_2 laser mixtures can also be predicted. This is shown (Lowke, Phelps, and Irwin, 1973) in Fig. 3.21. These are calculations as a function of the current density. The curves start off independent of current density, but eventually as the electric field goes up, the efficiency goes down. It appears that rather detailed predictions can be made regarding the CO_2 laser, although there may be a 20% discrepancy between the observed and the measured gain.

Let us now turn our attention to electronic transition lasers. Among the best understood are the molecular line lasers involving hydrogen, nitrogen, and perhaps someday, sulfur. The general picture is that the electrons excite the hydrogen molecule from the ground vibrational state to the $B^1\Sigma_u^+$ shown in Fig. 3.22 (Dreyfus and Hodgson, 1974). If irradiation

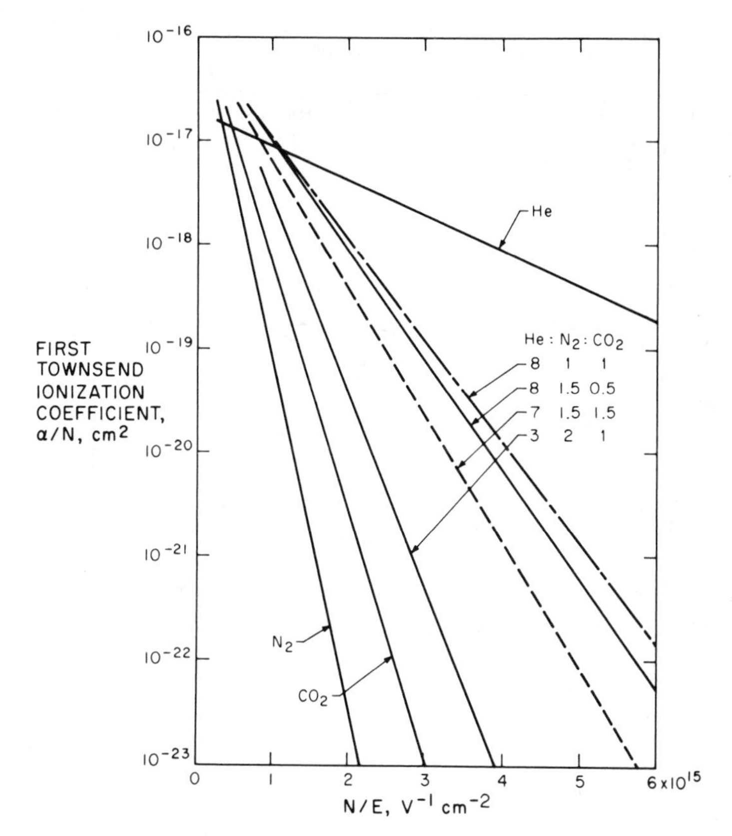

Fig. 3.19 Calculated Townsend ionization coefficients for various CO_2 mixtures.

takes place near the outer turning point, the molecules radiate into empty vibrational levels and population inversion is achieved. The question arises as to what information is available regarding electron collisions in hydrogen. Cross sections for the hydrogen molecule as a function of electron energy are shown in Fig. 3.23. The upper curve is the momentum transfer cross section. The lower curves show various inelastic processes including rotational excitation and threshold vibrational excitation. At the higher energies, rotational excitation is based on theory and over much of the energy range, theory was used to obtain cross sections for electronic excitation. There are some conflicting electron beam experiments and at higher electron energies there is poor agreement between the various theories. However, using this set of cross sections predictions can be made of the behavior of the hydrogen molecule. In Fig. 3.24 are plotted various rate coefficients as a function of N/E. The energy range is between 2 and

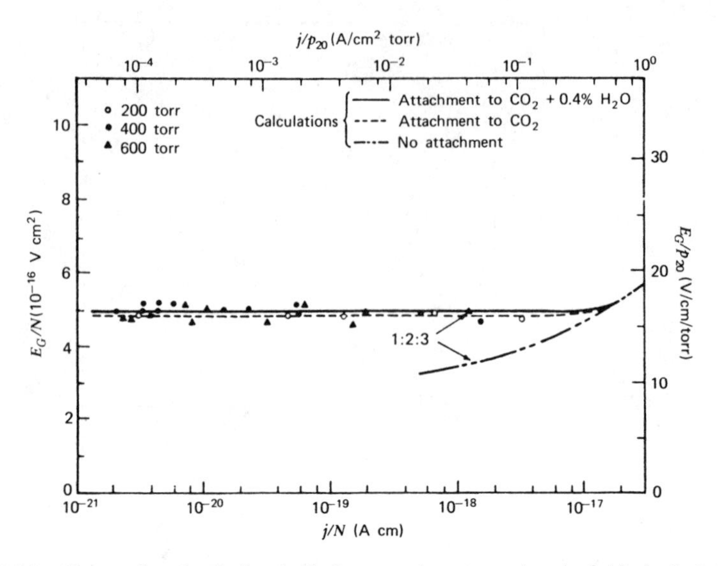

Fig. 3.20 Measured and calculated discharge maintenance electric fields in 1:2:3 mixtures of CO_2:N_2:He.

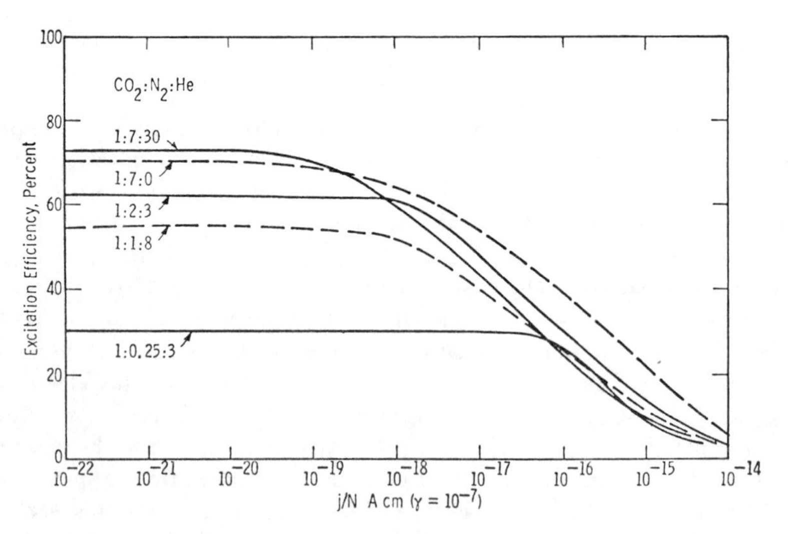

Fig. 3.21 Excitation efficiency for various laser mixtures in discharge excited CO_2 lasers.

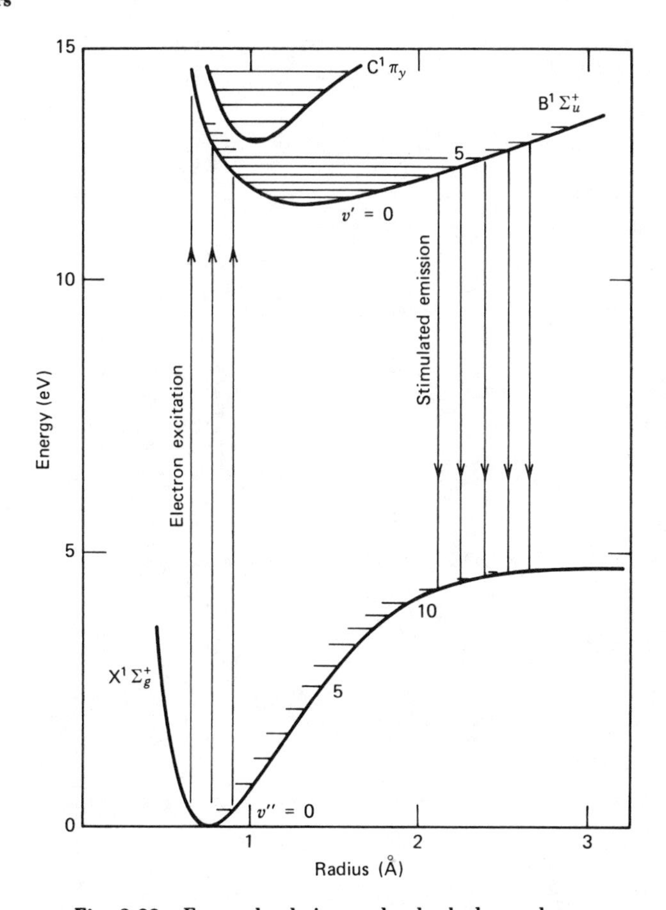

Fig. 3.22 Energy levels in a molecular hydrogen laser.

20 eV. The experimental data for the ionization coefficient are very accurately known but unfortunately there is not good agreement between theory and experiment. For the less accurately measured processes of u.v. excitation, the agreement appears rather good. In the case of dissociation, the experiments were done under difficult conditions and need to be done again.

A proposed molecule for electron impact excitation and subsequent lasing is the sulfur molecule. The potential energy curves for this molecule are shown in Fig. 3.25 (Leone and Kosnik, 1977). Unfortunately nothing is known about electron collisions in molecular sulfur.

A class of lasers that work well is the rare gas excimer laser. Xenon is an example, and the potential energy scheme for Xe_2 is shown in Fig. 3.26 (Rhodes, 1974). It can be seen that there are a number of atomic as well as

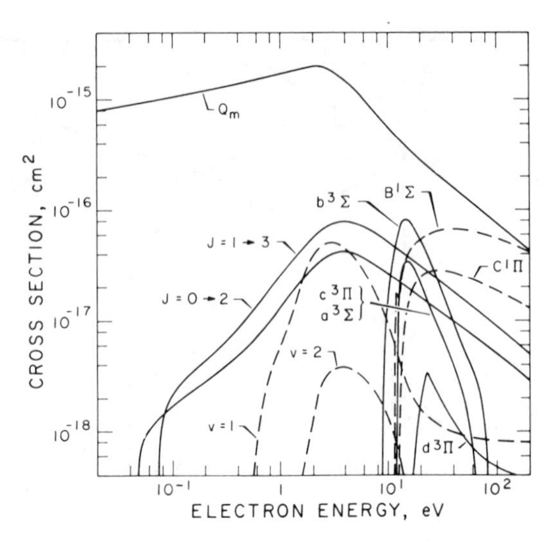

Fig. 3.23 Electron collisions with H_2.

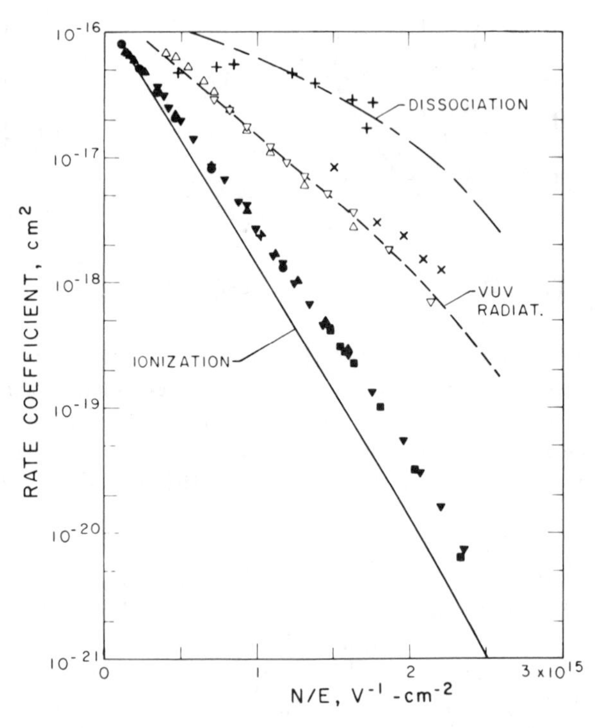

Fig. 3.24 Ionization and excitation coefficients for H_2.

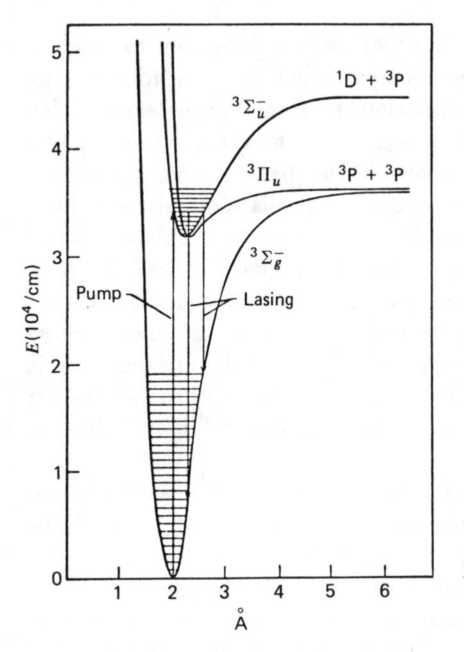

Fig. 3.25 Potential energy curves for S_2.

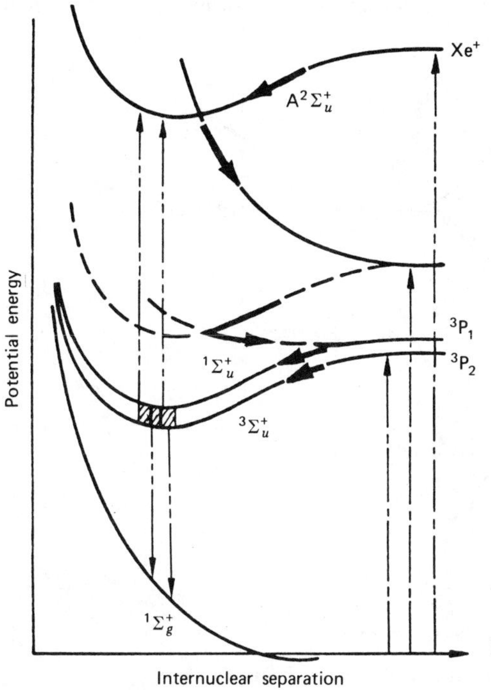

Fig. 3.26 Kinetics of a xenon laser.

molecular processes of importance. Electron impact excitation to produce the atomic ion usually results from bombardment by very high energy electron beams. These atomic ions associate to form the molecule which then undergoes dissociative recombination. How these atomic states find their way into the lower levels is unknown at the present time. Apparently they do so very rapidly, and are followed by molecular formation to form the ${}^1\Sigma_u^+$ and the ${}^3\Sigma_u^+$ levels which can then radiate to the repulsive ground state to give the laser action. The concept of the excimer is that this level dissociates very rapidly so that there is an absence of absorbers. Electron collisions also appear to be very important in coupling the lower energy levels and perhaps in such processes as dissociation of the molecules. It is unfortunate that the radiation from these lasers is too far in the ultraviolet for many applications. They are now being proposed as light bulbs for excitation of other molecules.

In contrast to the rare gas excimer lasers, the halogen based excimer lasers are finding many applications. One of the newest of these is the mercury chloride laser. The potential energy curves for this system are shown in Fig. 3.27 (Parks, 1977; Duzy, Hyman, 1977). The usual excitation scheme is by high energy electron beam and as such may not involve electron-molecule collisions. However, recent work suggests the use of a discharge excitation scheme where the mercury molecules are excited and then

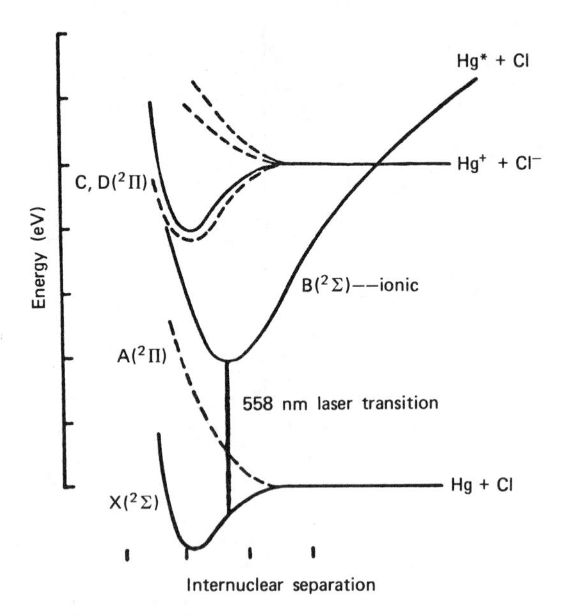

Fig. 3.27 Potential energy curves for HgCl.

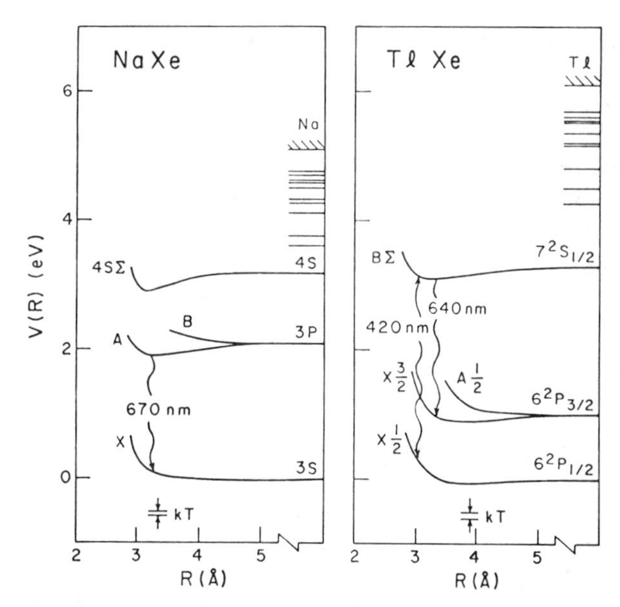

Fig. 3.28 Potential energy curves for NaXe and TlXe.

associate to form an excited HgCl molecule. These then radiate to the bound ground state. In terms of electron-molecule collision mechanisms, one of interest is the electron impact excitation of the ground state to the repulsive state. The threshold is quite small, around 1 eV, and therefore should occur at a rapid rate.

Finally, let us briefly discuss metal based excimer lasers, such as the NaXe and TlXe molecules shown in Fig. 3.28 (Gallagher, 1979). There are no obvious electron-molecule collisions in these diagrams. But if a detailed model is worked out, dissociative recombination of electrons and molecular ions turns out to be a crucial process and may be the major energy loss process in these systems.

In conclusion it should be emphasized that there are many applications of electron-molecule collision phenomena in current technology. Unfortunately, as is clear from this paper, each time we try to find a set of data for analysis of a specific scheme, we discover much that we do not have. A great deal of work still needs to be done in this field.

REFERENCES

Altschuler, S., 1957, *Phys. Rev.*, **107**, 114.

Bell, Alexis T., 1971, *Chem. Engr. Progress Symp. Ser.*, **67**, No. 112, 1.

Boness, M. J. W., and G. J. Schulz, 1974, *Phys. Rev.*, **A9**, 1969.

Cartwright, D. C., S. Trajmar, A. Chutjian, and W. Williams, 1977, *Phys. Rev.*, **A16**, 1041.

Chang, R. P. H., and A. K. Sinha, 1976, *Applied Phys. Lett.*, **29**, 56.

Chantry, P. J., 1969, *J. Chem. Phys.*, **51**, 3369.

Chen, C. L., and P. J. Chantry, 1978, *Bull. Am. Phys. Soc.*, **23**, 140.

Chutjian, A., D. C. Cartwright, and S. Trajmar, 1977, *Phys. Rev.*, **A16**, 1052.

Collins, L. A., and D. W. Norcross, 1977, *Phys. Rev. Lett.*, **38**, 1208.

Denes, L. J., and J. J. Lowke, 1973, *Appl. Phys. Lett.*, **23**, 130.

Dreyfus, R. W., and R. T. Hodgson, 1974, *Phys. Rev.*, **A9**, 2635.

Duzy, C., H. A. Hyman, *Chem. Phys. Lett.*, **52**, 345, (1977).

Gallagher, A. C., in *Excimer Lasers*, Ed. C. K. Rhodes, Springer-Verlag, 1979 (Vol. 30, Topics in Applied Physics, Chap. 5).

Geltman, S., and K. Takayanagi, 1966, *Phys. Rev.*, **143**, 25.

Hall, R. I., and S. Trajmar, 1975, *J. Phys.*, **B8**, L293.

Lawton, S. A., and A. V. Phelps, Feb. 1978, *J. Chem. Phys.*, **69**, 1055.

Leone, S. R., and K. G. Kosnik, 1977, *Appl. Phys. Lett.*, **30**, 346.

Linder, F., and H. Schmidt, 1971, *Zeit. f. Naturforsch.*, **26**, 1617.

Lowke, J. J., A. V. Phelps, and B. W. Irwin, 1973, *J. Appl. Phys.*, **44**, 4664.

McTaggart, F. K., 1967, *Plasma Chemistry in Electrical Discharges*, Amsterdam, pp. 36, 125.

Parks, J. H., 1977, *Appl. Phys. Lett.* **31**, 192.

Phelps, A. V., 1972, *Ann. Geophys.*, **28**, 611.

Phelps, A. V., unpublished.

Phelps, C. T., and R. F. Griffiths, 1976, *J. Appl. Phys.*, **47**, 2929.

Rees, M. H., and R. A. Jones, 1973, *Planet. Space Sci.*, **21**, 1213.

Rhodes, C. K., 1974, *IEEE J. of Quantum Electronics*, **QE-10**, 153.

Schulz, G. J., 1964, *Phys. Rev.*, **135**, A988.

Spence, D., and G. J. Schulz, 1972, *Phys. Rev.*, **A5**, 728.

Spencer, F. E. Jr., and A. V. Phelps, May 1976, *Proc. of the 15th Symp. on the Eng. Aspects of MHD*, Philadelphia, Paper IX.9.1.

Swanson, N., and R. J. Celotta, 1975, *Phys. Rev. Lett.*, **35**, 783.

Trajmar, S., D. C. Cartwright, and W. Williams, 1971, *Phys. Rev.*, **A4**, 1482.

Trajmar, S., W. Williams, and W. Kupperman, 1972, *J. Chem. Phys.*, **56**, 3759.

Wong, S. F., M. J. W. Boness, and G. J. Schulz, 1973, *Phys. Rev. Lett.*, **31**, 969.

CHAPTER 4

Laboratory Experiments

FRANZ LINDER

Department of Physics
University of Kaiserslautern
Kaiserslautern, Federal Republic of Germany

1 INTRODUCTION

The last 15 years have been a period of great activity and very fruitful work in electron-atom and electron-molecule scattering. It is adequate to mark the beginning of this period by two important laboratory experiments. Both were performed by George Schulz.

Figure 4.1 shows the famous cross sections for vibrational excitation in e-N_2 scattering (Schulz, 1962, 1964). Since these cross sections were measured by Schulz in 1962, this has become a famous model case in electron-molecule scattering. It was the first experiment that clearly demonstrated that a resonance process was involved, i.e., the temporary formation of a short-lived N_2^- ion. There was evidence before, but this was

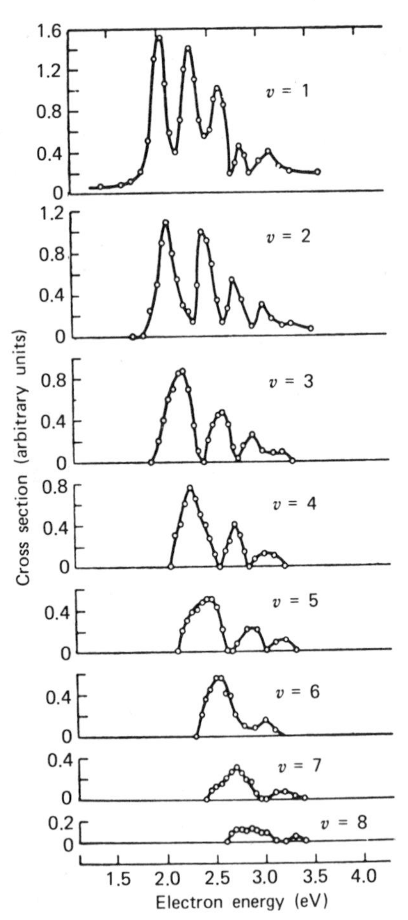

Fig. 4.1 Energy dependence of the vibrational excitation in N_2 by electron impact. The angle of observation was 72°. Excitation of up to eight vibrational states of N_2 is visible in the region of the $N_2^-\ ^2\Pi_g$ compound state. Vibrational excitation is very small outside the resonance region. (From Schulz, 1962, 1964.)

a clear and direct experiment. The experiment shows that the formation of short-lived intermediate states (a short-lived N_2^- in the present case) represents a very effective mechanism for the transfer of electron energy into molecular vibration. The cross sections for vibrational excitation are very large inside and essentially zero outside the resonance region. This observation established the dominant role of resonances in low-energy electron-molecule scattering.

Although the concept of a resonance process was immediately clear, the problem turned out to be rather demanding for theorists. Eventually, a satisfactory explanation was given by the equally famous boomerang model (Herzenberg, 1968; Birtwistle and Herzenberg, 1971) shown in Fig. 4.2. The incident electron is attached to the molecule at an internuclear distance R_0 and an N_2^- is formed. The nuclear wave packet moves out and bounces back, which means that the molecular bond stretches and contracts. During this movement of the nuclei, the resonant state decays again by emitting an electron and leaving the N_2 in a vibrationally excited state.

The second important experiment is shown in Fig. 4.3. It was the discovery of a narrow resonance in e-He scattering (Schulz, 1963) at 19.3 eV. This is the other famous observation that marks the beginning of this field. We have the case of an isolated resonance in a single channel problem, a most beautiful example of a resonance.

The apparatus used for these experiments is shown in Fig. 4.4 (Schulz, 1962, 1963, 1964). Electrons are produced by a hot filament and energy-selected by a 127° electrostatic energy selector. The electron beam of variable energy is crossed with an atomic or molecular beam and controlled by an electron collector. Scattered electrons are analyzed by a second 127° energy analyzer and detected by a multiplier. It is fascinating to see how simple the original apparatus was. Nowaday's instruments are much more sophisticated, but this first instrument contained all the essential features which made the experiments possible.

This chapter is mainly on electron-molecule scattering, but let me spend a few paragraphs more on the e-He resonance. The e-He resonance is a classical experiment and as such it is represented in the teaching laboratory* in Kaiserslautern. The apparatus is shown in Fig. 4.5. An electron gun produces a beam of monoenergetic electrons which is crossed with a gas beam and controlled by a monitor. The scattered electrons are detected at variable angle by a simple detector system. Everything has to be rather simple, since the instrument has been built for the teaching laboratory.

A typical result obtained by the students is shown in Fig. 4.6. The figure shows the e-He resonance measured at scattering angles between 20°

*Teaching laboratory for third year students.

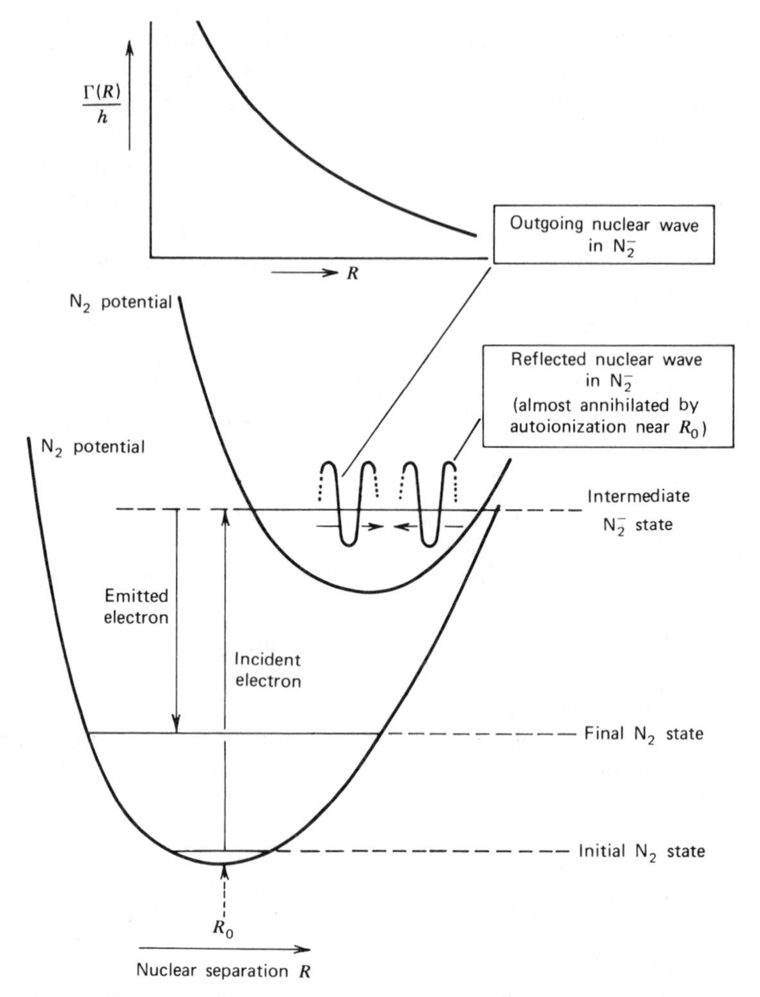

Fig. 4.2 The "boomerang" model of the nuclear wavefunction applied to the N_2^- ion. This model is discussed by Herzenberg (1968). It is based on the assumption that the magnitude and R-dependence of the width $\Gamma(R)$ are such that only a single outgoing and a single reflected wave are important. (From Herzenberg, 1968; Birtwistle and Herzenberg, 1971.)

and 138°. This result is analyzed by the students in the following way. At energies around 19.3 eV, we have a superposition of direct scattering ($e + \text{He} \rightarrow e + \text{He}$) and resonant scattering ($e + \text{He} \rightarrow \text{He}^- \rightarrow e + \text{He}$). One writes down a partial wave analysis for the scattering amplitude

$$f(\vartheta, E) = \frac{1}{2ik} \sum_{l=0}^{2} (2l+1)(e^{2i\delta_l} - 1) \cdot P_l(\cos\vartheta) \tag{4.1}$$

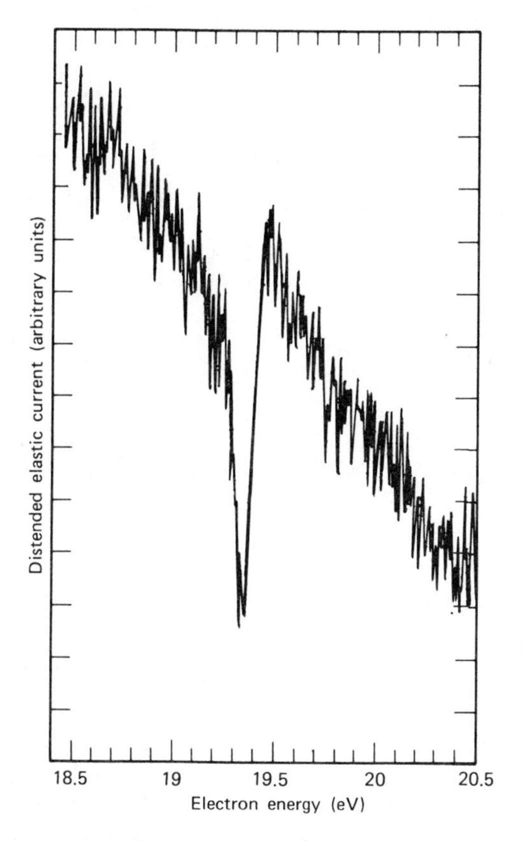

Fig. 4.3 The original observation of the $(1s2s^2)^2$ S resonance in helium. The angle of observation is 72° and the elastic cross section is measured. The decrease in the cross section near 19.3 eV is approximately 14%. (From Schulz, 1963.)

where it is sufficient to consider three contributing partial waves $l = 0, 1, 2$. The resonance occurs in the s-wave, so the corresponding phase shift δ_0 is written as a sum of a background phase shift δ_{bg} and a resonance phase shift δ_{res}:

$$\delta_0 = \delta_{bg} + \delta_{res} \tag{4.2}$$

where $\delta_{bg} \approx \text{const}$ and δ_{res} shows a rapid variation with energy according to

$$\delta_{res} = -\arctan \frac{\Gamma/2}{E - E_R} \tag{4.3}$$

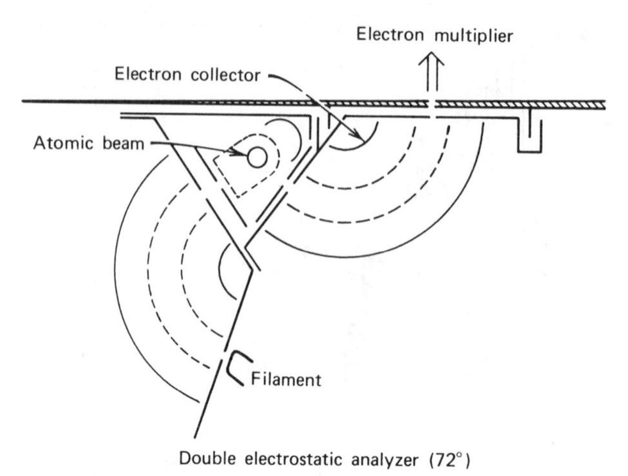

Fig. 4.4 Schematic diagram of the double electrostatic analyzer, the original instrument used by Schulz. (From Schulz, 1963.)

Fig. 4.5 Schematic diagram of the electron scattering apparatus used in the teaching laboratory of the physics department in Kaiserslautern to measure the e-He resonance.

It rises from 0 to π over an energy interval given by Γ, the width of the resonance. Finally, by using the relation

$$\frac{d\delta}{d\Omega}(\vartheta, E) = |f(\vartheta, E)|^2 \tag{4.4}$$

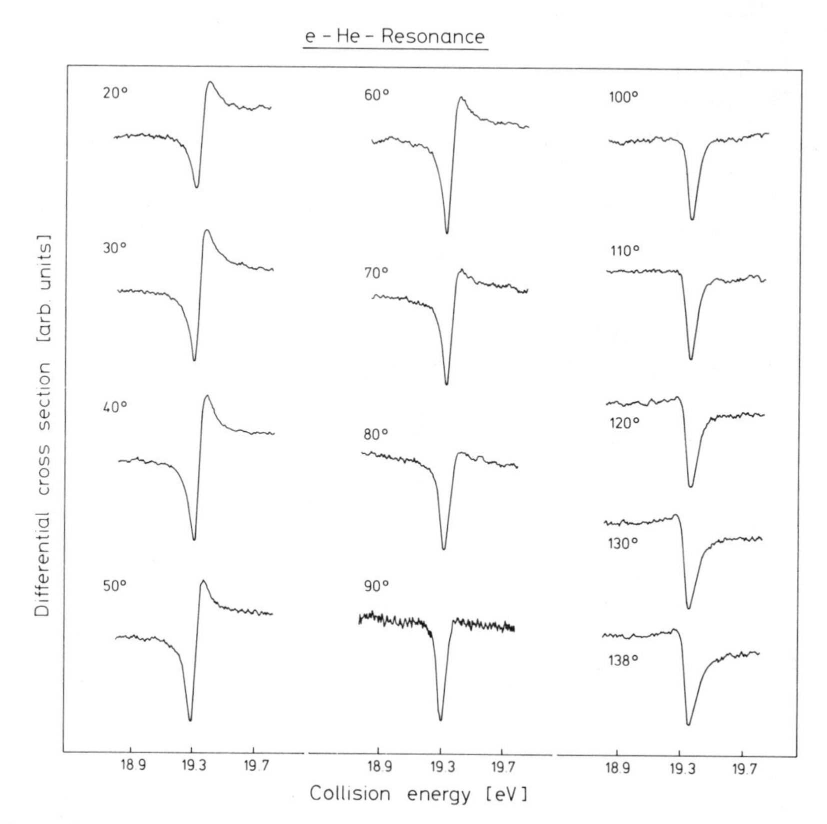

Fig. 4.6 Typical experimental results obtained by the students using the apparatus of Fig. 4.5.

For the differential cross section, the student is able to analyze the experimental results.

A list of classical experiments represented in the teaching laboratory is shown in Fig. 4.7 (altogether there are about 25 experiments). The $e-\mathrm{He}$ resonance is one of them and the last questionnaire among students showed that it is one of the most interesting ones.

The experimental technique has been refined over the years. The $e-\mathrm{He}$ resonance measured by a research apparatus is shown in Fig. 4.8 (Andrick, Langhans, Linder, and Seng, 1975). In addition to the 19.3 eV resonance, which appears as an enormous effect on this scale, one observes cusp structures at the opening of the inelastic $e-\mathrm{He}$ channels. There is a lot of beauty in clear physical measurements.

Let me conclude these introductory remarks. The two experiments mentioned above are classical experiments. An important aspect of labora-

1. Stern–Gerlach Experiment
2. Millikan Experiment
3. Mass Spectrometer
4. Rutherford Scattering
5. Mössbauer Effect
6. Zeeman Effect
7. e – He-Resonance
8. Optical Pumping
9. Microwave Experiment
10. X-rays (Structure Analysis)

Fig. 4.7 List of classical experiments represented in the teaching laboratory at Kaiserslautern. The list continues to about 25 experiments.

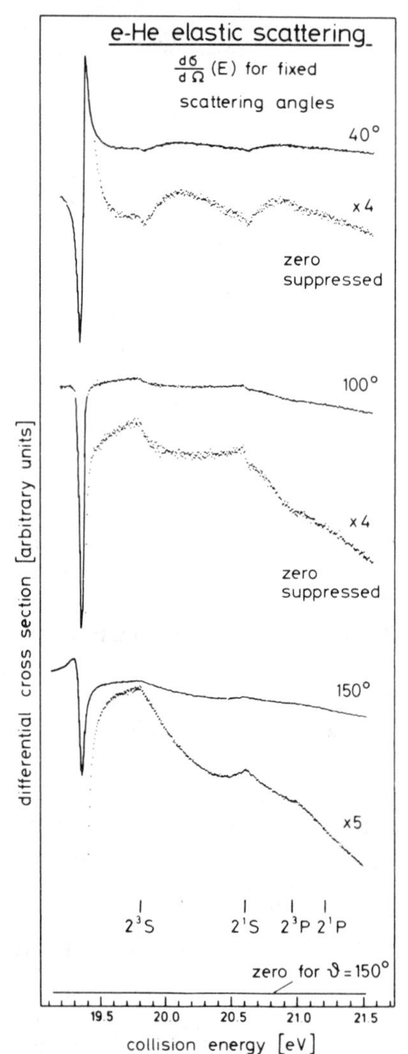

Fig. 4.8 Measurements of the differential cross section for elastic e-He scattering in the vicinity of the $n = 2$ states. In addition to the 19.3 eV resonance, cusp structure is observed at the opening of the inelastic channels. (From Andrick, Langhans, Linder, and Seng, 1975.)

tory experiments is always to put emphasis on model cases. The interaction between theory and laboratory experiment provides us with basic models that are important for a general understanding of the processes. The above two model cases have played this role and they have opened a whole field.

In the meantime, many resonances have been observed in electron-atom and electron-molecule scattering (Schulz, 1973a,b). The dominating role of resonances for all processes at low energies is well established (Schulz, 1976). A description of the processes with neglect of resonances would be meaningless. One works on the classification of types of resonances, on the determination of energies, widths and lifetimes, branching ratios for the decay into various channels, etc., and there has been a great activity in this respect during the last 15 years.

2 RECENT DEVELOPMENTS

After the first impact to a field and the development to a certain state, further work can go in different ways. It can be in the direction of refinement on an existing line or it can concentrate on search for new possibilities. Both lines are certainly important and it is more a matter of personal taste and attitude that is preferred. In the following, some examples to illustrate recent developments in this field have been selected. Depending on the case, they are more on one line or the other.

2.1 Spectroscopy of Dissociative Attachment and Vibrational Excitation

The first example illustrates recent progress in the spectroscopy of dissociative attachment (DA) and vibrational excitation (VE) obtained by Hall (1977) and coworkers at the University of Paris. The achievements of this group consist in refined studies of DA processes combined with high sensitivity measurements for VE processes. Hereby, the close relationship that exists between VE and DA becomes directly visible in their experiments.

Let us assume a model case. In Fig. 4.9 we have two potential curves of the neutral molecule XY and some negative ion states. We want to focus our attention on the dissociating XY^- curve. By attaching an electron of the appropriate energy, the molecule forms an XY^- state (point E, R_0 in Fig. 4.9). The XY^- ion starts to dissociate along its repulsive potential curve. All along this way, the system decays into a free electron and a neutral molecule in vibrationally excited states, until it reaches the crossing point of the XY^- and XY curves, from where it is stabilized

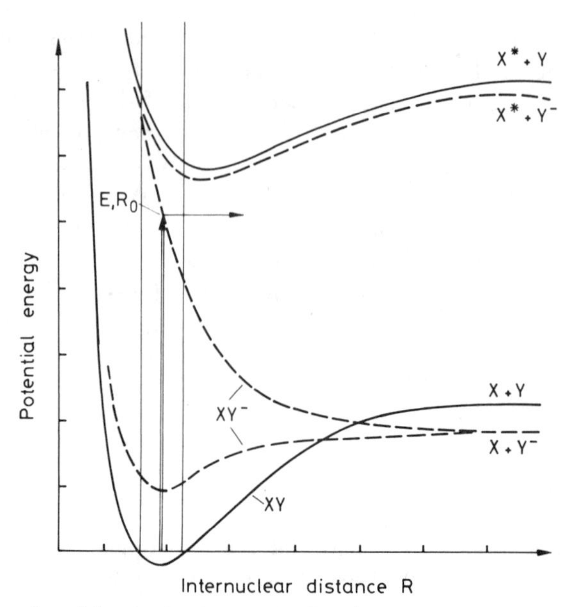

Fig. 4.9 Assumed model case showing potential curves of the neutral molecule XY (full lines) and some negative ion states XY^- (dashed curves). The model case is directly applicable to the H_2 molecule.

against this decay channel and it continues to dissociate into $X + Y^-$. From this picture, one derives the close relationship between DA and VE which in principle, of course, is well known.

The model case of Fig. 4.9 directly applies to a real case, the H_2 molecule. Out of the variety of results that have recently been obtained by the Paris group, let us pick out one example, the spectroscopy of DA and VE processes processes when 10 eV electrons are captured by an H_2 molecule.

Before discussing the results, let us show the apparatus in Fig. 4.10 (Hall, Cadez, Schermann, and Tronc, 1977). Energy-selected electrons are produced by an electron gun and crossed with a gas beam. A rotatable analyzer detects scattered electrons and negative ions produced by DA. As a special feature, the analyzer has a momentum filter based on a weak magnetic field. With the filter off, both electrons and ions are detected, whereas with the filter on, only the ions can pass to the multiplier. The group uses a second instrument of a similar kind that is not described here. The advantages in this work are the following: the energy and angular distributions of both the electrons and the negative ions can be measured, the experiments combine a good resolution in energy and angle with a high sensitivity, and both processes (VE and DA) can be studied in

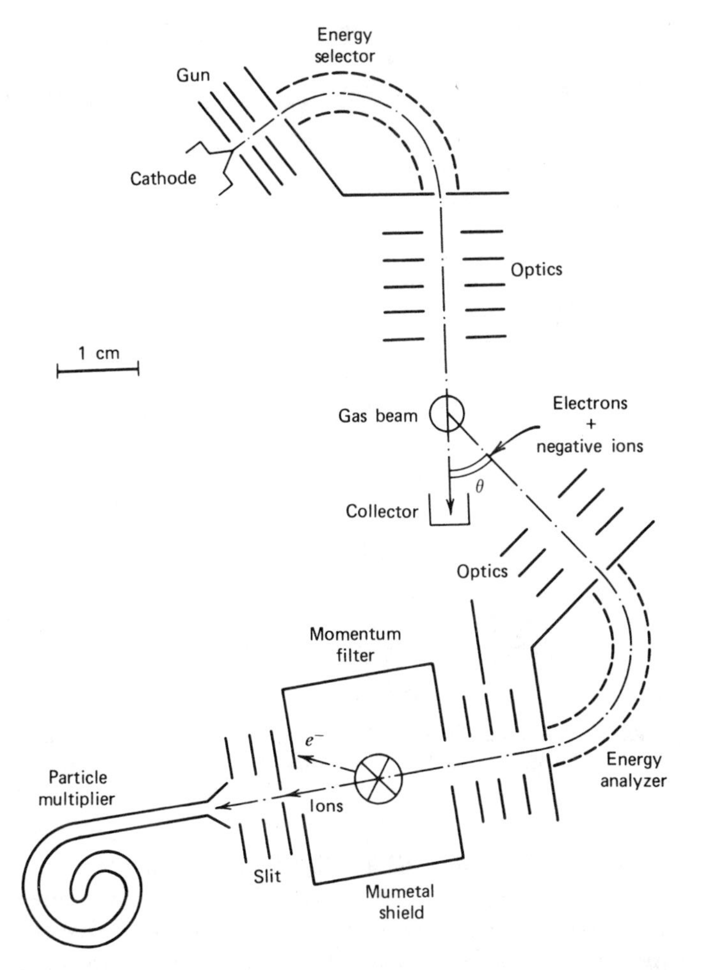

Fig. 4.10 Schematic diagram of the apparatus used by Hall and co-workers for the measurement of vibrational excitation and dissociative attachment. (From Hall, Cadez, Schermann, and Tronc, 1977.)

the same instrument which guarantees a most direct connection between the investigations.

Some of the results are shown in the next figures. Figure 4.11 shows the energy dependence of H^- formation between 10 and 13 eV at an angle of 50° (Tronc, Fiquet-Fayard, Schermann, and Hall, 1977). The broad maximum centered at 10.5 eV is attributed to the $H_2^{-2}\ \Sigma_g^+$ repulsive state that crosses the Franck–Condon region at about this energy. The fine structure observed between 11 and 13 eV is interpreted as predissociation of the

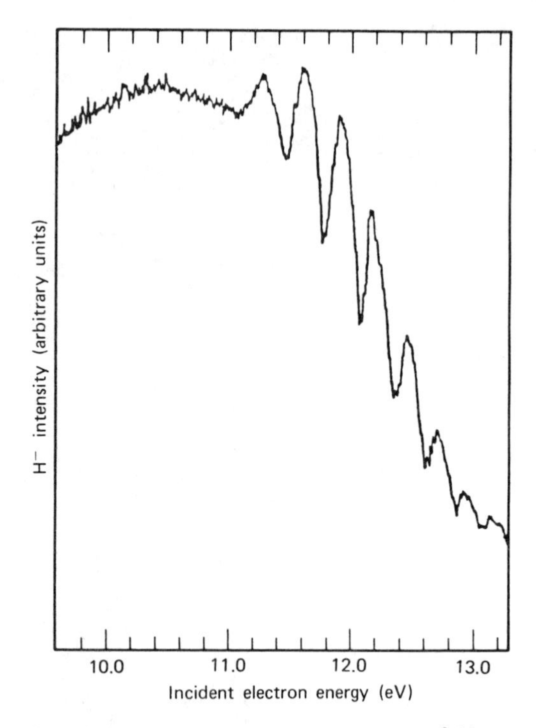

Fig. 4.11 Energy dependence of H^- formation from H_2 via the $^2\Sigma_g^+$ resonance between 10 and 13 eV at a 50° scattering angle. (From Tronc, Fiquet-Fayard, Schermann, and Hall, 1977.)

vibrational levels of a higher bound H_2^- state into the continuum of the repulsive state (see Fig. 4.9). Such a fine structure had already been observed in a total collection experiment by Dowell and Sharp (1968), but much more refined studies are possible in a differential scattering experiment.

Figure 4.12 gives measured angular distributions of H^- in the 9–13 eV electron energy range (Tronc, Fiquet-Fayard, Schermann, and Hall, 1977). The full curves represent a theoretical fit to the measured points. A $^2\Sigma_g^+$ symmetry has been assumed for the resonant state and two partial waves ($l = 0$ and 2) have been included in the partial wave analysis.

The main reason for including this experiment in this chapter is given by the following result (Hall, 1977). The combined spectroscopy of DA and VE allows one to trace out the decay of the resonant state in a very detailed manner. Figure 4.13 shows energy loss spectra for scattered electrons in H_2 and D_2 at 10 eV. Vibrational excitation of the target molecule is observed up to the dissociation limit. A detailed interpretation

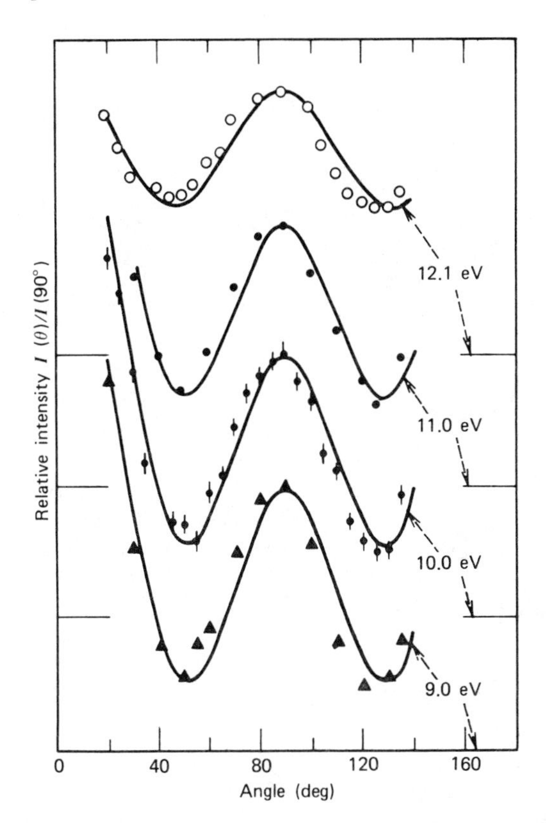

Fig. 4.12 Measured angular distributions of H^- from H_2 at the indicated electron energies (E_i). The values $E_i = 9$, 10, and 11 eV correspond to the structureless part of the dissociation continuum shown in Fig. 4.11, but $E_i = 12.2$ eV corresponds to the $v = 3$ level of a bound H_2^- state (series a) predissociated by the same continuum. The full curves represent the best fits obtained using a theoretical model discussed by Tronc and co-workers. (From Tronc, Fiquet-Fayard, Schermann, and Hall, 1977.)

can be given by Fig. 4.14. The XY^- molecular ion is formed at $E = 10$ eV and an internuclear distance R_0. Starting at this point, the nuclei begin to separate along the repulsive potential curve. The energy which is entirely electronic at the beginning is gradually transferred into nuclear kinetic energy (indicated by arrows in Fig. 4.14). The decay of the resonant state obeys the Franck–Condon principle, i.e., the transitions are vertical and the nuclear kinetic energy accumulated by the XY^- system during its lifetime is conserved. For example, if the decay occurs at R' or R'', the XY molecule is found in vibrationally excited states v' or v'', respectively. For decay at R_d, the nuclear kinetic energy is sufficient to reach the dissociation limit $X + Y$. Therefore, the decay between R_0 and R_d leads to

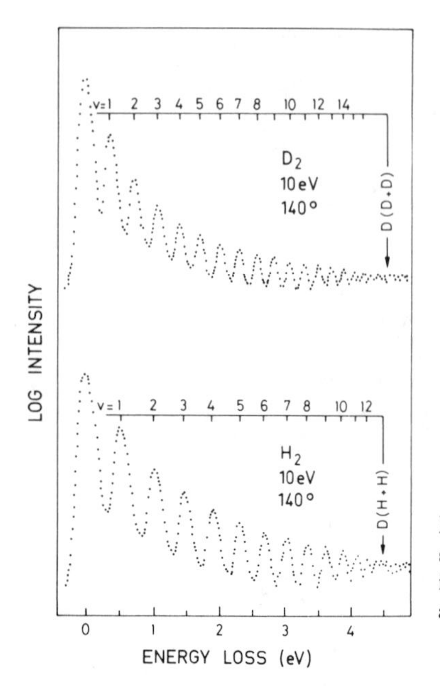

Fig. 4.13 Energy-loss spectra for electron scattering from H_2 and D_2 at 10 eV and 140°. The intensity is plotted on a logarithmic scale (see also Fig. 4.15). (From Hall, 1977.)

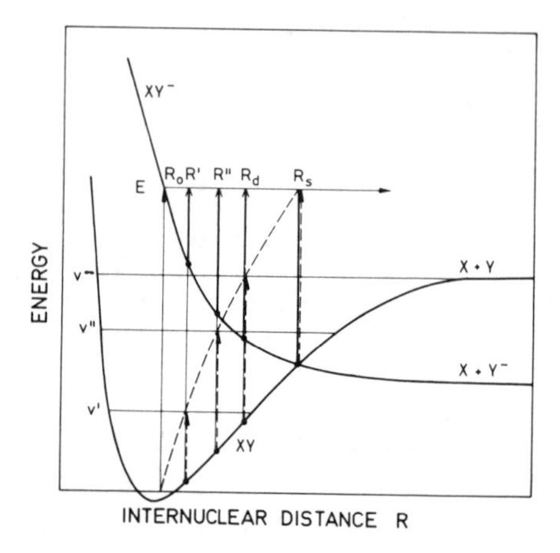

Fig. 4.14 Schematic diagram representing transitions between the dissociating state XY^- and high vibrational levels of XY. (From Hall, 1977.)

vibrational excitation $XY(v) + e$, the decay between R_d and R_s gives dissociation into neutral fragments $X + Y + e$, whereas after R_s the XY^- system cannot decay by emission of an electron any more and it dissociates into $X + Y^-$ (dissociative attachment). All this is observed in the present measurements and it illustrates the possibilities of this spectroscopy.

Figure 4.15 shows a plot of the peak intensities of vibrational excitation versus the vibrational energy transferred. A logarithmic scale is used

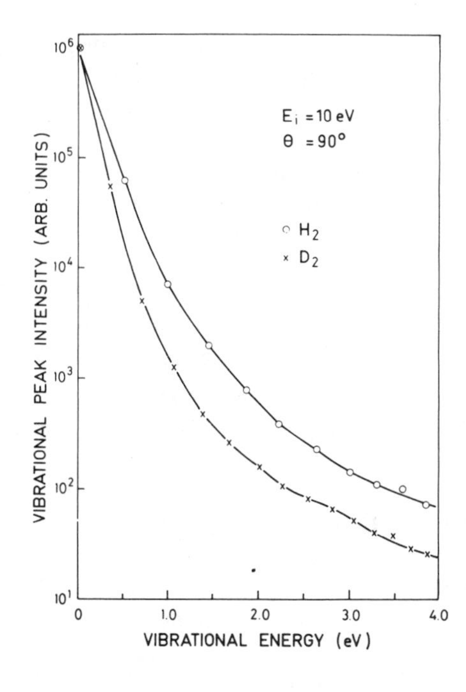

Fig. 4.15 Vibrational excitation intensities in H_2 and D_2 at 10 eV and 90°. The elastic peaks are both normalized to 10^6. (From Hall, 1977.)

demonstrating the intensity decrease over five orders of magnitude which also illustrates the sensitivity of the measurements. The results show an isotope effect. When normalized at $v = 0$, the intensities decrease faster for D_2. The reason is that the D_2^- separates more slowly, the decay into low vibrational levels is preferred and less intensity is left for the higher vibrational levels. Also the survival probability for DA is lower, leading to a smaller cross section for D^-/D_2 than for H^-/H_2 in accordance with earlier measurements (Rapp, Sharp, and Briglia, 1965).

The refined measurements in this work do not give really surprising results in this simple case. What is appealing is the fact that we have a very clear experiment for a simple model case and the connection between DA and VE becomes directly visible. The application of the method to more complicated cases which is under way in the Paris laboratory is expected to open new possibilities for the study of fragmentation processes.

2.2 Electron-Polar Molecule Scattering

As the next topic I want to say a few words on electron-polar molecule scattering and discuss some of the results we have obtained in our experiments at Kaiserslautern (Linder, 1977). I want to include this topic here, since the observed phenomena are fairly striking and we expect these phenomena to be of general importance, when electron scattering is extended to larger molecules in future studies. The emphasis in our work was on the range of very low collision energies (≤1 eV). It is in this energy range where the phenomena are observed, and I believe that the interaction between electrons and molecules, especially polyatomic molecules, at these very low energies will be a field of great importance in the future.

In Fig. 4.16 we have a table of selected examples of polar molecules divided into groups according to the magnitude of their dipole moment. For the present aspects, CO and NO can be regarded as essentially nonpolar molecules. The molecules LiF, etc., represent examples of strongly polar molecules of ionic character. We have performed measurements for the molecules of the second group which possess dipole moments in the vicinity of the so-called critical dipole moment $D_0 = 1.625$ debye, the minimum value for a stationary dipole to bind an electron.

Figure 4.17 shows energy-loss spectra (Rohr and Linder, 1976; Rohr, 1977, to be published) for HF, HCl and HBr in order to illustrate the processes that have been investigated. Vibrational excitation is clearly separated, whereas rotational excitation is not resolved and always included. We will concentrate on the discussion of vibrational excitation (VE).

Measured excitation functions for VE are shown in Figs. 4.18–4.21 (Rohr and Linder, 1976; Rohr, 1977, to be published). All cross sections are dominated by resonance scattering, the cross sections are too large by far to be explained by a direct process. The angular dependences are essentially isotropic in the whole energy range shown. The most striking

Molecule	Dipole Moment (Debye)
CO	0.10
NO	0.15
HF	1.82
HCl	1.11
HBr	0.83
H_2O	1.85
LiF	6.5
CsF	7.9
NaCl	9.5
KI	10.8

Fig. 4.16 Selected examples of polar molecules. (From Linder, 1977.)

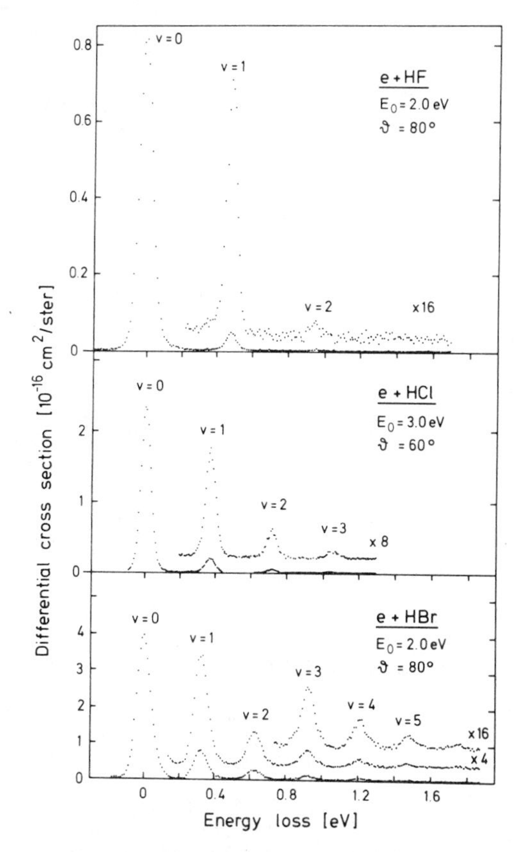

Fig. 4.17 Energy-loss spectra for HF, HCl, and HBr at the collision energies and scattering angles indicated in the figure. (From Linder, 1977.)

feature is a strong and sharp threshold peak in all cases. The general structure of the excitation functions for these molecules seems to consist of a broad resonance region on which a relatively sharp threshold peak is superimposed. The details depend on the individual case, but the general structure is the same for all of these molecules. It should also be mentioned that these are extremely large cross sections for VE by electron impact. In the case of HBr, for example, the cross sections are larger by a factor of 10 than the corresponding cross sections in N_2.

An interpretation of these results can be given by the following model. HCl is taken as a representative case. Figure 4.22 shows the potential curve of the neutral HCl molecule (full line) and two proposed potential curves for HCl^- (dashed lines). In this picture, VE proceeds by the following mechanism. The incoming electron (say with an energy of around

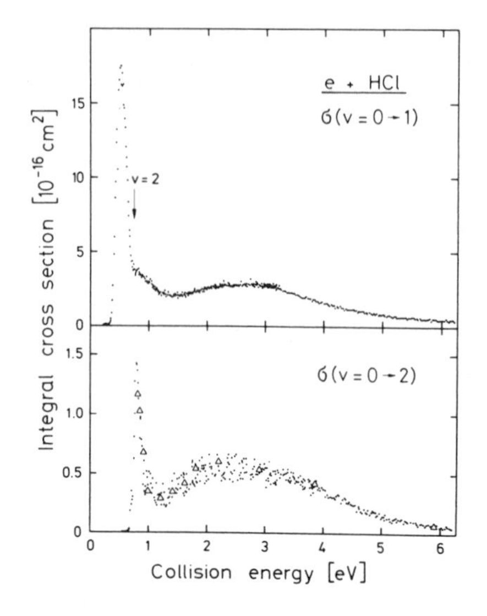

Fig. 4.18 Integral cross sections for vibrational excitation in HCl. (From Linder, 1977; Rohr and Linder, 1976.)

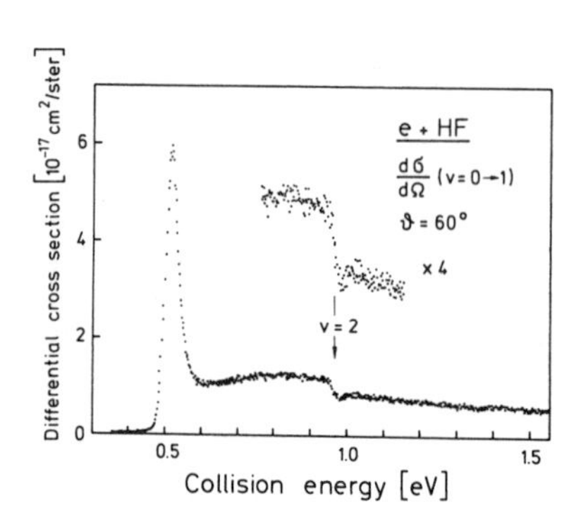

Fig. 4.19 Differential excitation function for $v = 1$ excitation in HF at $\vartheta = 60°$. (From Linder, 1977; Rohr and Linder, 1976.)

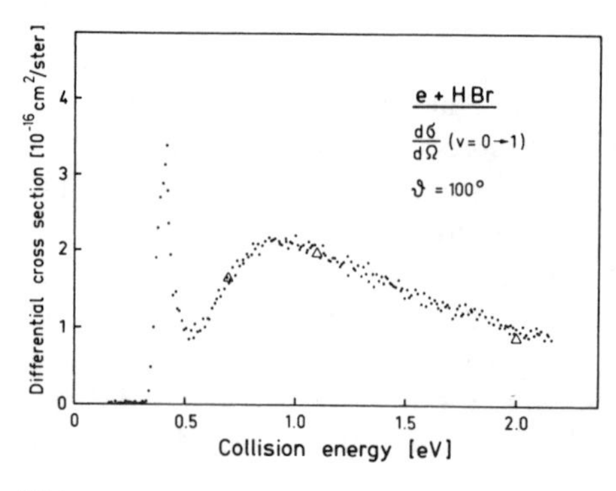

Fig. 4.20 Differential excitation function for $v = 1$ excitation in HBr at $\vartheta = 100°$. (From Linder, 1977; Rohr, 1977.)

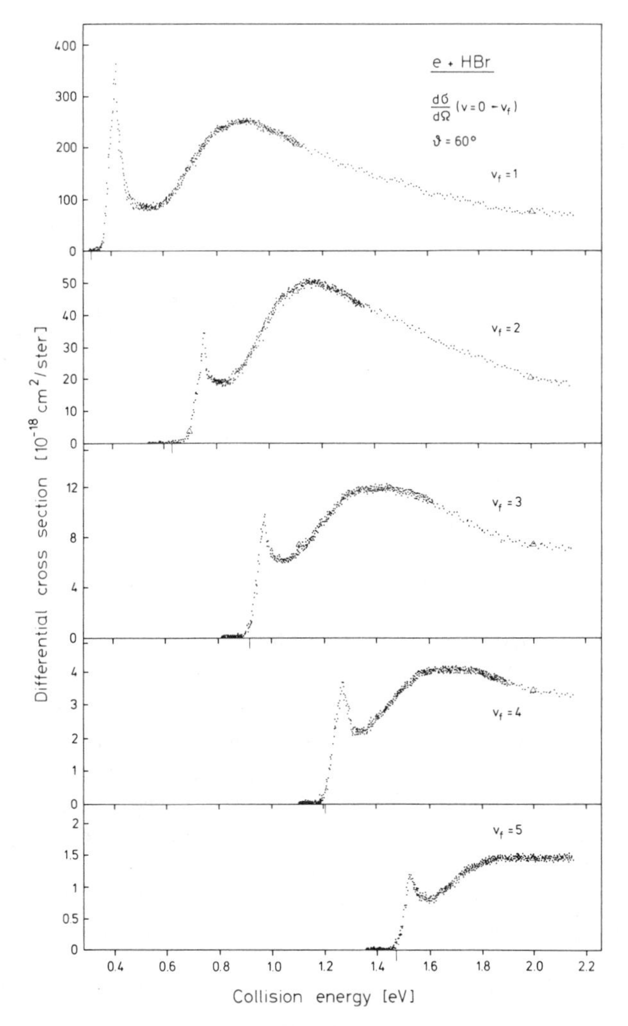

Fig. 4.21 Differential excitation functions for vibrational excitation to $v_f = 1$ to 5 in HBr at $\vartheta = 60°$. The angular dependence of vibrational excitation is isotropic in the whole energy region shown. (From Rohr, 1977.)

1 or 2 eV) "feels" the repulsive branch of the first HCl^- state (the one that dissociates into $H + Cl^-$). This causes a broad enhancement of the VE cross sections at these energies. At the threshold of a vibrational channel, we get an additional effect. The very slow outgoing electron is trapped in the long range dipole field of the molecule causing a resonant state of the outgoing electron. This resonant state is represented by the second HCl^-

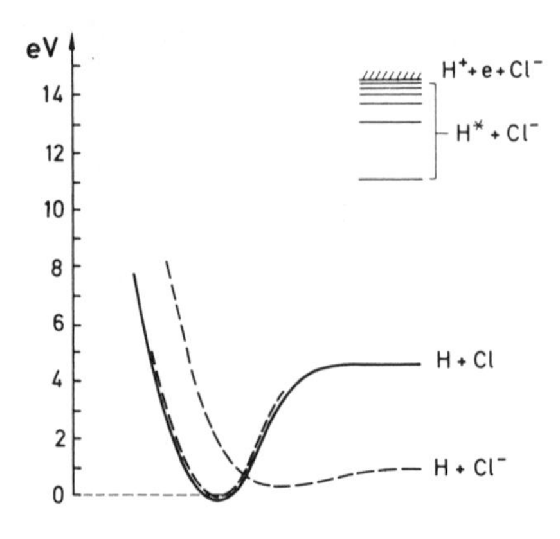

Fig. 4.22 Proposed potential curves of the HCl^- system (dashed lines). The full line represents the HCl ground state. (From Linder, 1977.)

curve in Fig. 4.22 that is drawn as nearly coincident with the potential curve of the neutral molecule. By this mechanism, we get an additional enhancement of the VE cross section at threshold which is due to the final state interaction. It has been shown by Dubé and Herzenberg (1977) that the magnitude of the threshold cross sections can be explained by this mechanism in a quantitative way (see Fig. 4.23).

A few more words about the nature of this second resonant state should be said. The essential ideas are provided by the dipole model* (see Fig. 4.24) (Wallis, Herman, and Milnes, 1960; Crawford, 1967; Turner, Anderson, and Fox, 1968). The polar molecule is represented by a finite dipole built up from point charges and fixed in space. For dipole moments less than the critical value $D_0 = 1.625$ debye, there is no bound state for an electron in this field. As soon as the dipole moment exceeds this critical value, one obtains an infinite number of bound states which are correlated with $H^* + q^-$ in the dissociation limit. This would correspond to $H^* + Cl^-$ in the case of Fig. 4.22.

It is interesting to look at the charge density of the extra electron in these states. Contour maps of the charge density are shown in Fig. 4.25 for two examples (Turner, Anderson, and Fox, 1968). For a large separation of the point charges corresponding to a large dipole moment, the charge density is concentrated around the positive point charge, i.e., the positive end of the dipole. This situation corresponds to an H atom and a negative point charge q^- sitting nearby. For a small separation of the point charges, we have the situation of the other example in Fig. 4.25 (note the difference in the z and ρ scales). In this case, the dipole moment is just above the critical value D_0 ($0.84\ ea_0$ corresponds to 2.13 debye). We have a bound

*The interesting history of this model has been described by J. E. Turner (1977).

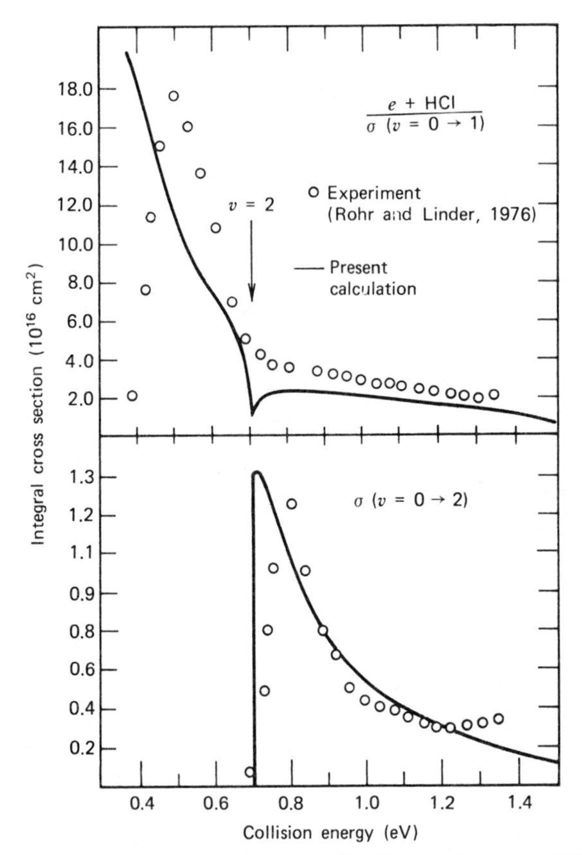

Fig. 4.23 Vibrational excitation cross sections for electron impact on HCl. Comparison between experiment (Rohr and Linder, 1976) and a theoretical model calculation by Dubé and Herzenberg based on the final state interaction. (From Dubé and Herzenberg, 1977.)

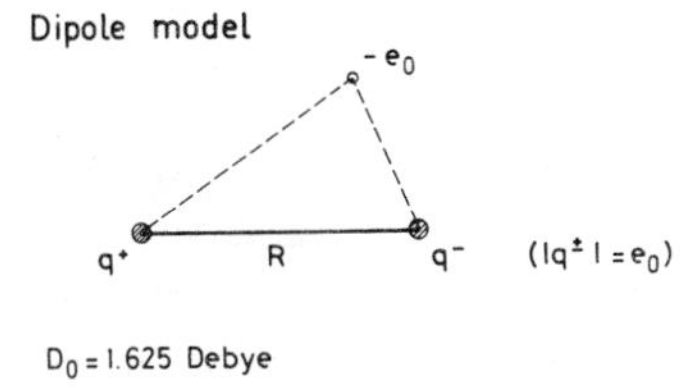

Fig. 4.24 Dipole model; the electron of charge $-e_0$ moves in the field of two fixed point charges q^+ and q^- separated by a distance R and having charges $+e_0$ and $-e_0$. The critical value for the occurrence of bound states is $D_0 = 1.625$ debye. (See Wallis, Herman, and Milnes, 1960; Crawford, 1967; Turner, Anderson, and Fox, 1968; Turner, 1977.)

state of the extra electron, but very weakly bound and with a very diffuse orbital.

The increase of the orbit of the extra electron as we approach the critical dipole moment D_0 is underlined by Fig. 4.26. This figure gives the mean distance of the extra electron from the positive end of the dipole

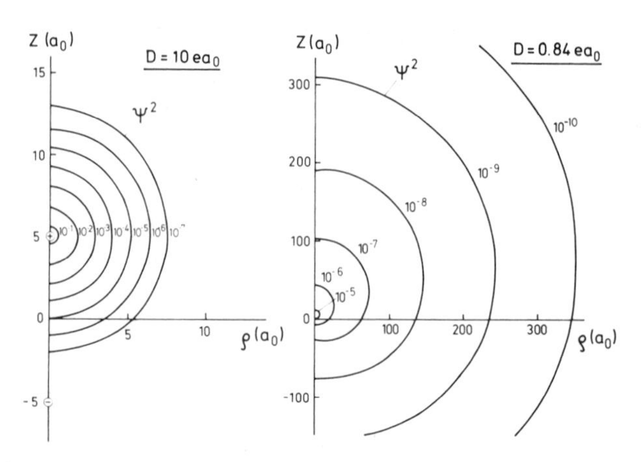

Fig. 4.25 Contour maps of the probability density of the electron bound in the field of the dipole. The dipole is directed along the z-axis and has a separation of the point charges of $10a_0$ and $0.84a_0$, respectively. (From Turner, Anderson, and Fox, 1968.)

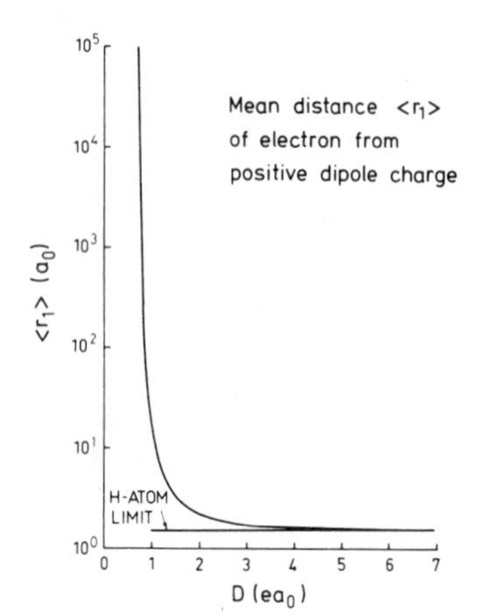

Fig. 4.26 Mean distance $\langle r_1 \rangle$ of the electron from the positive dipole charge $q^+ = +e_0$ as a function of the dipole moment D. (From Turner, Anderson, and Fox, 1968.)

(Turner, Anderson, and Fox, 1968). Coming from the H-atom limit at large D, we obtain a dramatic increase of the orbit as we approach D_0. The present molecules have dipole moments in the vicinity of D_0 and we expect very diffuse orbits for the negative ion states of these systems.

It is necessary, of course, to introduce some modifications and refinements to this simple model when applying it to a real molecule. The internal structure of the molecule has to be taken into account and the real molecule is a rotating and vibrating system. However, the essential points and the general value of the simple model are confirmed by more refined treatments and *ab initio* calculations (Crawford and Garrett, 1977).

Before leaving this topic, let us look again at the HF case. Jordan and Wendoloski (1977) have calculated the binding energy of the extra electron by *ab initio* methods. For the stationary molecule, they obtain a positive electron affinity in the order of 10^{-5} eV. The electron is in a very diffuse orbit of extremely low binding energy. The resulting HF^- is a very large and very fragile system. It is almost certain that the Born–Oppenheimer approximation is not applicable for such a system and that the coupling between electronic and nuclear motion plays an essential role. Looking at the experimental result (see Fig. 4.19), the threshold effects are seen to be very narrow. The observed width of approximately 40 meV is mainly instrumental. It is obvious that HF is an extremely interesting model case both experimentally and theoretically.

It is necessary to understand these model cases in detail. They are probably of general importance when scattering experiments are extended to larger (polyatomic) molecules many of which have appreciable dipole moments, quadrupole moments, etc. A wealth of interesting phenomena can be expected in these molecules, particularly in the very low energy range.

2.3 Energy Resolution—The State of the Art

The problem of energy resolution is of general interest in electron-molecule scattering. This concerns the experimental resolution of rotational transitions as well as the spectroscopy of resonances. Most of the results in electron-molecule scattering have been obtained with instruments using electrostatic devices as energy selectors. The types mostly used are the 127° cylindrical selector and the 180° spherical monochromator. Magnetic fields are rarely used. A famous exception is the trochoidal monochromator developed at Yale.

The discussion is divided in the following way. First, a few words are said on the current state of standard type instruments and the energy resolution available with these instruments. Then, some recent developments using lasers are discussed.

A typical standard type instrument is shown in Fig. 4.27. This is an example from the Kaiserslautern laboratory and shows the apparatus used for the electron-polar molecule experiments (Rohr and Linder, 1976). The individual features of the instruments vary from laboratory to laboratory, but the general characteristics are always the same. The progress in energy resolution has not been dramatic since the early work of Schulz (1962, 1963, 1964) and of Simpson (1964) and Kuyatt and Simpson (1967). The present possibilities are approximately the same in several laboratories around the world. As a figure of merit for an electron gun one usually

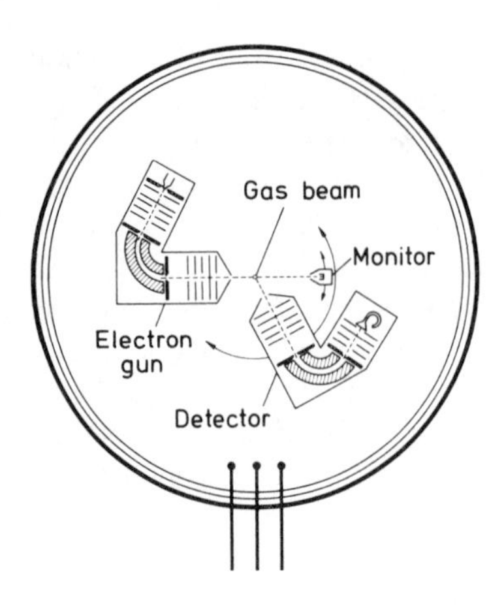

Fig. 4.27 Schematic diagram of a typical electron scattering apparatus. The detector can be rotated in a plane perpendicular to the gas beam. (From Rohr and Linder, 1976.)

considers the quantity $I(\Delta E)$, i.e., the electron current I available at a given energy resolution ΔE. As an example for the present possibilities, some recent results from the Kaiserslautern group are shown (Rohr and Linder, 1976), simply because they are best available to me. Figure 4.28 shows these $I(\Delta E)$ data obtained with the electron gun of Fig. 4.27. The best resolution obtained under reasonable operating conditions was $\Delta E = 7$ meV at a current of 5×10^{-10} A.

The best overall conditions, in my opinion, are presently available in the apparatus of Wong at Yale. As an example, Fig. 4.29 shows energy-loss spectra for H_2 and D_2 measured with this apparatus (Chang and Wong, 1977). The energy resolution in these spectra is 11.5 meV (FWHM of the energy-loss peak). This value contains contributions from the gun and the detector system. An estimate for the energy width of the gun alone gives approximately $\Delta E = 8$ meV. Under these conditions, the rotational transitions in the H_2 spectrum are completely resolved and Chang and Wong (1977) were able to study isotope effects in rotational and vibrational excitation by electron impact. So the criteria for a good instrument are not only a good energy resolution, but also a reasonable current and good operation conditions (i.e., stability, etc.).

What is the possibility of extending these measurements to other molecules? The energy resolution necessary for measuring rotational excitation in N_2, for example, can be illustrated by Fig. 4.30. This figure gives simulated energy-loss spectra for $e - N_2$ scattering. Buckley (1977) has calculated the cross sections for all relevant transitions and superimposed the energy-loss peaks with an assumed instrumental shape. The resulting spectra are shown in Fig. 4.30, where two different temperatures of the N_2 gas and an instrumental resolution of 5 and 10 meV have been assumed. Only with an energy resolution of 5 meV, one is able to separate

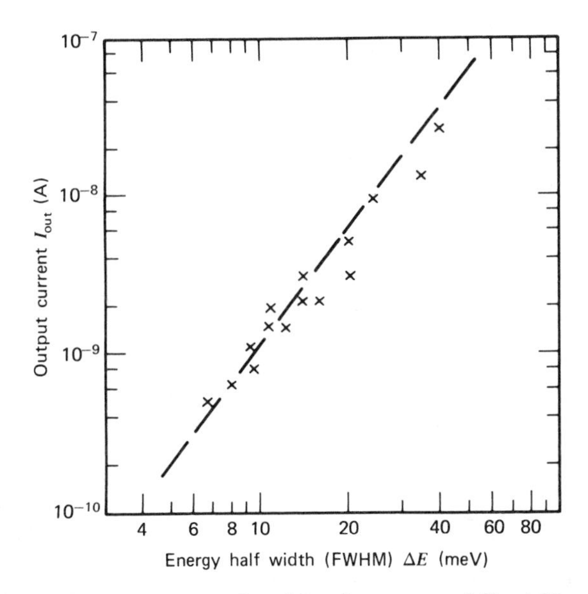

Fig. 4.28 Measured output currents I_{out} of the electron gun of Fig. 4.27 as a function of the energy width ΔE (FWHM). The crosses are the experimental results and the broken line represents an interpolation with slope 5/2 according to the expected law $I_{out} = c \cdot (\Delta E)^{5/2}$ for space-charge limited current. Such a plot is considered as a figure of merit for an electron gun. (From Rohr and Linder, 1976.)

the bulk of rotational transitions from elastic scattering, but one is still far from being able to resolve individual rotational transitions. So the situation is not hopeless, but also not very satisfying.

It is clear, therefore, that one is urged to search for new possibilities. One approach which could promise a larger step in improving the energy resolution is undertaken by Gallagher and coworkers at JILA in Boulder (Van Brunt and Gallagher, 1977; Gallagher and York, 1974). The idea is to use photoelectrons as a monoenergetic source of electrons. The principle of the approach is shown in Fig. 4.31. A barium beam containing metastables Ba* is crossed with a laser beam and the photoelectrons are extracted in a direction perpendicular to both beams. The full apparatus is sketched in Fig. 4.32. The insert shows a level diagram of Ba indicating that the $^1D_2$ metastables are photoionized by the laser line and 17 meV photoelectrons are produced. A portion of these photoelectrons is extracted by a lens system and formed to a beam which is then scattered off a target gas beam produced by a nozzle-skimmer arrangement. The scattered electrons are detected by a multiplier.

In order to test the energy resolution obtained by this arrangement the photoelectrons are scattered from an Ar beam and the well-known Ar

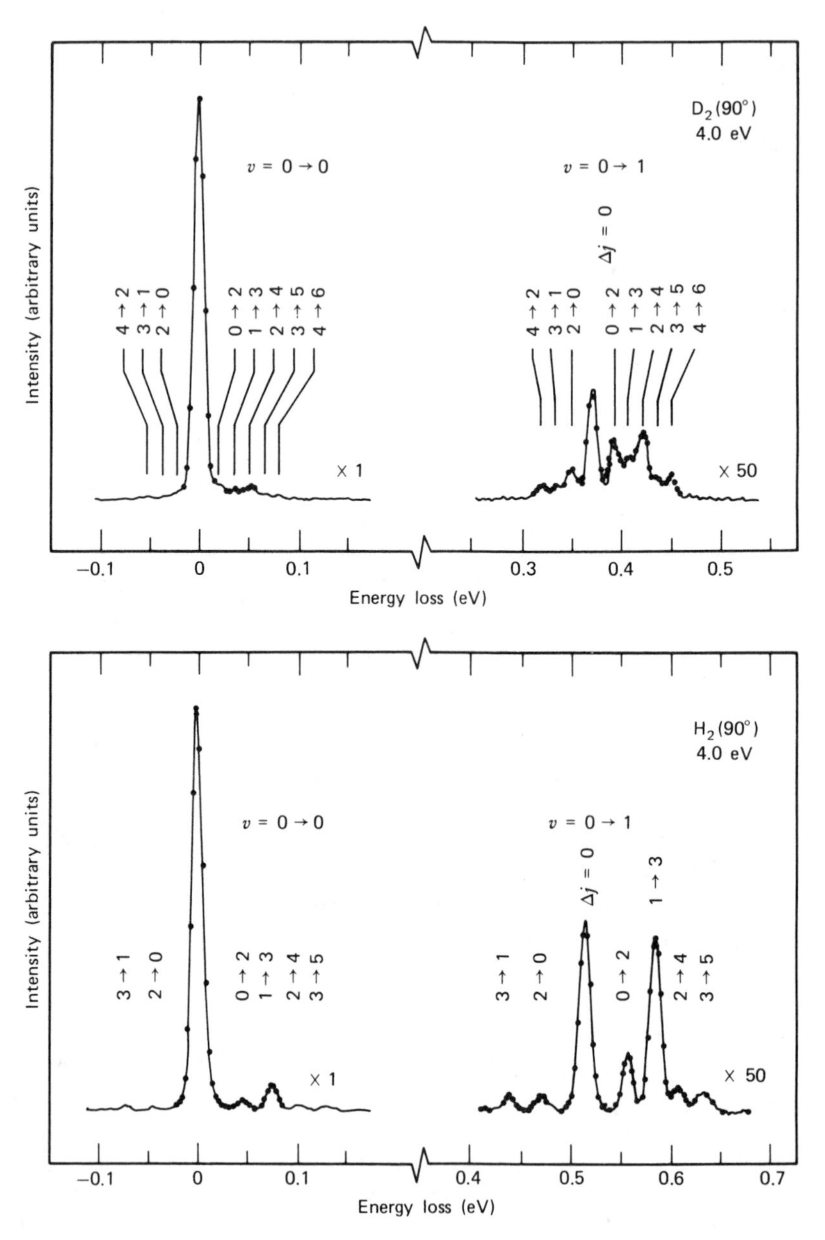

Fig. 4.29 Energy-loss spectra for H_2 and D_2 at 4 eV and 90°. The energy resolution is 11.5 meV (FWHM in the energy-loss peaks). Vibrational and rotational transitions are identified. (From Chang and Wong, 1977.)

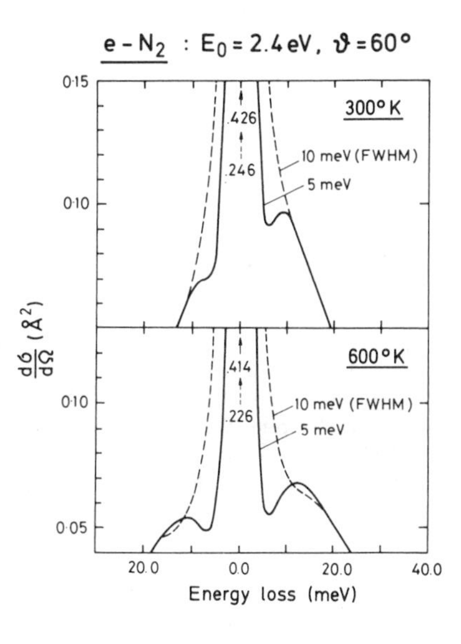

Fig. 4.30 Simulated energy-loss spectra for e-N_2 scattering at 2.4 eV and 60°. The intensities of all relevant transitions are calculated and superimposed with an assumed instrumental peak shape. Two different gas temperatures (300 and 600°K) and two different energy resolutions (5 and 10 eV) are considered. (From Buckley, 1977.)

resonances near 11 eV are measured. The result is shown in Fig. 4.33 (Van Brunt and Gallagher, 1977b). The measured points are from different runs and the solid line represents a best fit to the experimental points. This best fit is obtained in the following way. By assuming a natural width $\Gamma = 3.5$ meV for the resonance and by using the elastic scattering phase shifts known from other experiments, one can construct the natural line shape of the resonance. This line shape is convoluted with a Gaussian energy smearing of width w and the best fit is obtained for $w = 2.2$ meV.

This value w represents the total experimental width which is composed of the source width ΔE, Doppler broadening in $e-\mathrm{Ar}$ scattering and possible contributions from electric fields in the collision region and energy drifts. The latter contributions are hard to estimate, whereas the Doppler effect contributes to about 0.6 meV. This means that the energy width of the source must be better than $\Delta E = 2$ meV.

It is interesting to discuss briefly the various factors limiting the source width ΔE in the present arrangement. The line width of the ionizing radiation (laser) is negligible. Doppler broadening of the photoelectrons decreases with the energy of the electrons and is quite small for the 17 meV electrons in the present case (about 0.1 meV). The influence of electric fields in the source region (due to surface effects) is minimized by using a small source volume far from the surfaces. The collection field from the lens system (with a collection efficiency of about 10%) contributes about 0.3 meV to the energy spread. The essential limitation for the energy width is given by the space charge of the residual Ba^+ ions in the

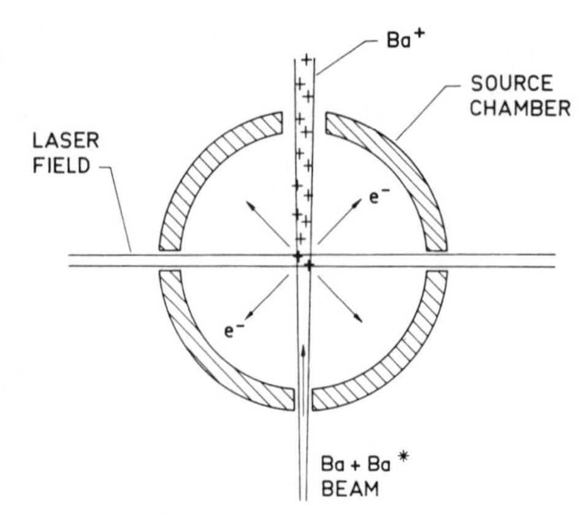

Fig. 4.31 Principle of the photoionization source to produce a beam of monoenergetic electrons. (From Van Brunt and Gallagher, 1977.)

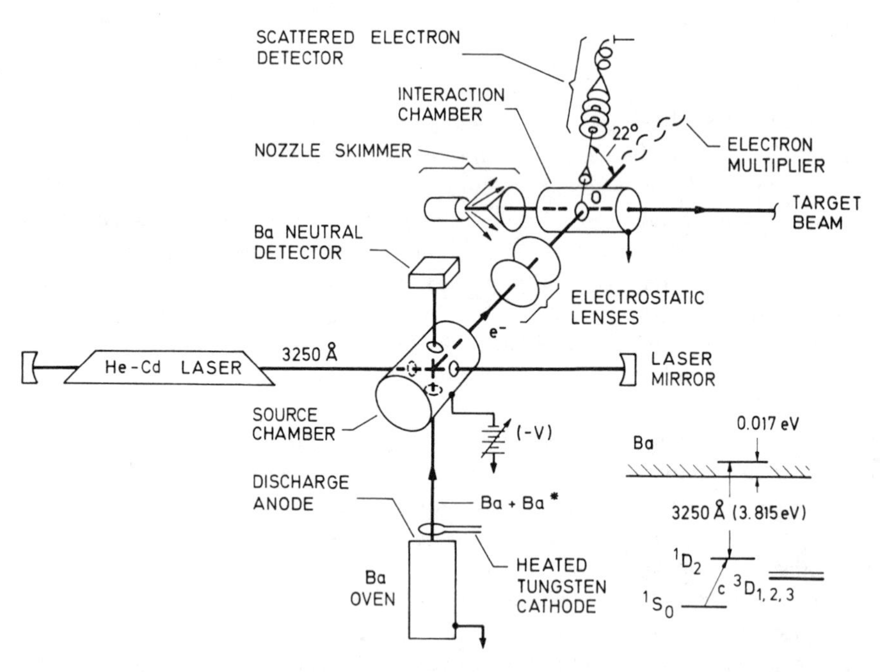

Fig. 4.32 Experimental arrangement of the electron scattering apparatus based on the photoionization source. The insert shows an energy level diagram of Ba indicating the relevant excitation and ionization transitions for electron production. (From Van Brunt and Gallagher, 1977.)

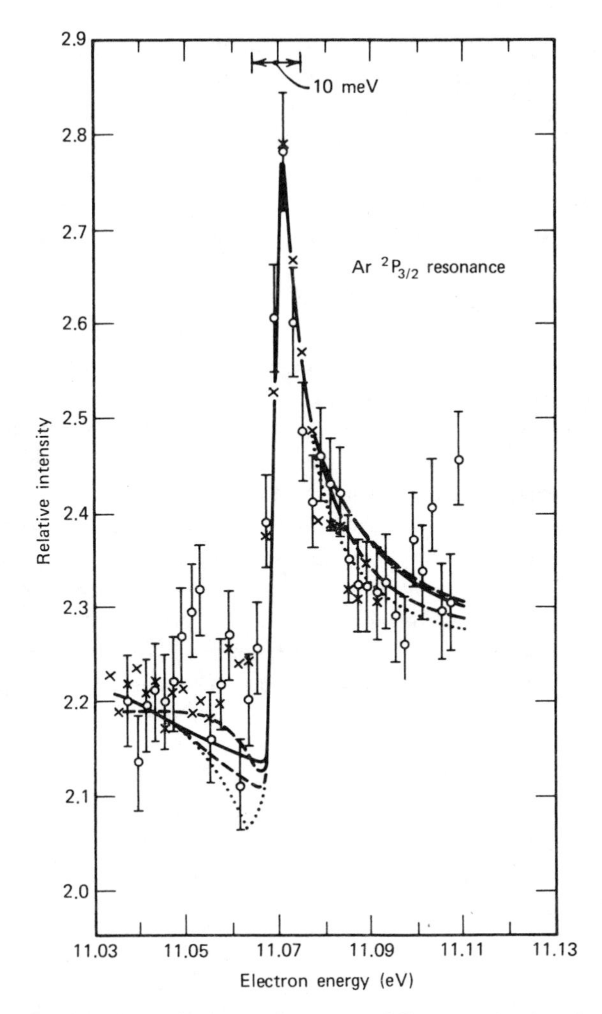

Fig. 4.33 Test result obtained with the apparatus of Fig. 4.32 showing the $(3p^5 4s^2)^2P_{3/2}$ resonance in elastic scattering from argon at 22°. The measured points are from different runs and the solid line represents a best fit to the experimental points. (From Van Brunt and Gallagher, 1977.)

source region, whereas the so-called "anomalous" energy spread (due to electron-electron interaction) is estimated to be small compared to the other effects.

The effect of primary importance in a photoionization source of monoenergetic electrons is the space charge of the residual ions. This leads to a direct proportionality between the current I and the energy width ΔE, since the ion density is proportional to the photoionization rate and hence the electron current. The present situation is depicted in Fig. 4.34 where the beam current I and the energy width ΔE are plotted on a double-logarithmic scale (Van Brunt and Gallagher, 1977a). The space charge limited current of the present photoionization source ($I \sim \Delta E$) is represented by the full line. The dashed line represents the $I \sim \Delta E^{5/2}$ law for the space charge limited current of an electrostatic selector. The proportionality factor in each law depends on the individual conditions of the source so that a parallel shift of each curve is always possible. In principle, however, there is always a crossing point of both curves and below a certain ΔE the photoionization source is better. The actual experimental results are also shown in Fig. 4.34. For the present photoionization source, the data point lies at $\Delta E = 2$ meV (or better) and $I = 8 \times 10^{-12}$ A. The crosses give the data points obtained by Rohr and Linder (1976) with a 127° selector (see Figs. 4.27 and 4.28).

We will now try to draw a conclusion. This experiment demonstrates that the approach of a photoionization source for monoenergetic electrons is realizable. The experiment is technologically rather complicated and it took about eight years of development to the present state. The method has not yet produced actually new results and the hope at the present time still rests on the traditional instruments as the general tool. However, not

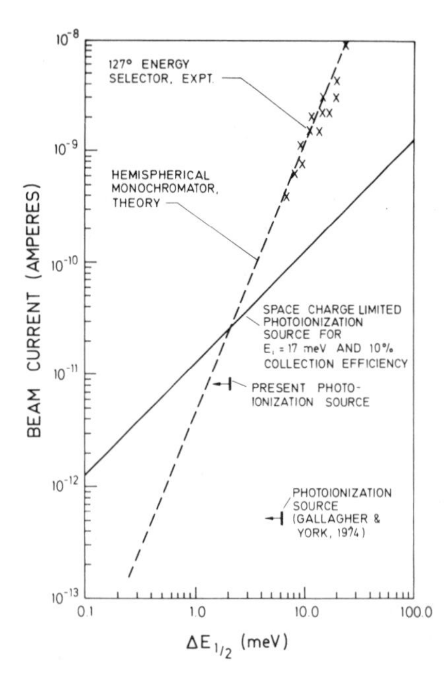

Fig. 4.34 Beam current versus electron energy spread for electrostatic energy selectors and photoionization source. The solid curve represents upper limits on current predicted for the present photoionization source where $I \sim \Delta E_{1/2}$. Experimentally obtained results are also indicated. The dashed curve represents a theoretical limit of the form $I \sim (\Delta E_{1/2})^{5/2}$ for an electrostatic energy selector (Simpson, 1964; Kuyatt and Simpson, 1967). The crosses are experimental data of Rohr and Linder (1976). (From Van Brunt and Gallagher, 1977.)

all potentialities of this source have been fully explored yet and there may well be a need for such sources in special applications.

As a final example in this context, I want to mention the experiment of Langendam, Gavrila, Kaandorp, and van der Wiel (1976) at the FOM-Institute in Amsterdam. This is also an experiment in the state of its infancy. The principle can be illustrated by Fig. 4.35 which shows a level diagram for Ne and Ne^-. Electrons of a suitable energy form a Ne^- state (say at 16.85 eV). A tunable laser can be used to induce an optical transition to a higher Ne^- state (at 18.95 eV) which now can decay into an excited Ne state (in the present case $^3P_1$ $(3p)$]. The formation of this state is detected by the Ne *I* emission at 5882 Å and 6030 Å.

The experimental set-up is shown in Fig. 4.36. In a collision chamber containing Ne gas, an intense electron beam and a laser beam of tunable wavelength are crossed. Photons emitted from the intersection volume of the two beams are detected by a photomultiplier after traversing an interference filter. A gated counting scheme serves to obtain a good signal to background ratio.

Figure 4.37 shows one of the first results of this experiment. In the

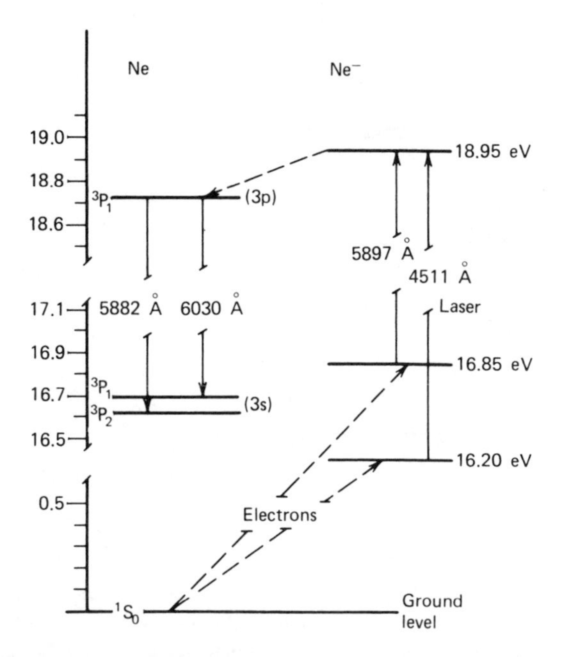

Fig. 4.35 The Ne and Ne^- energy levels involved in the Langendam and van der Wiel experiment. The Ne^- levels are from Schulz (1973). (From Langendam, Gavrila, Kaandorp, and van der Wiel, 1976.)

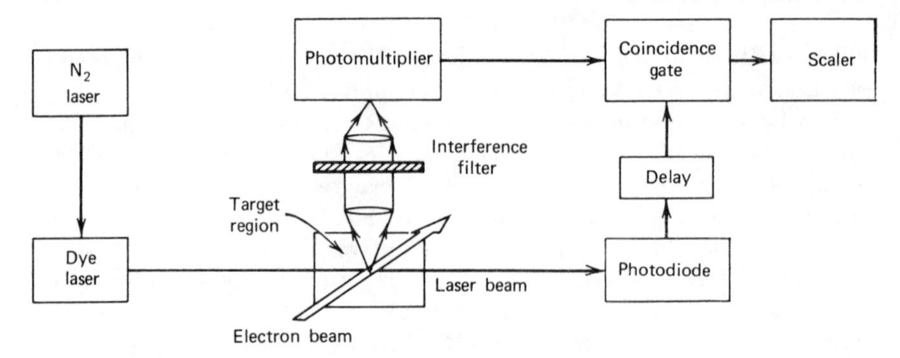

Fig. 4.36 Experimental arrangement of the Langendam and van der Wiel experiment. (From Langendam, Gavrila, Kaandorp, and van der Wiel, 1976.)

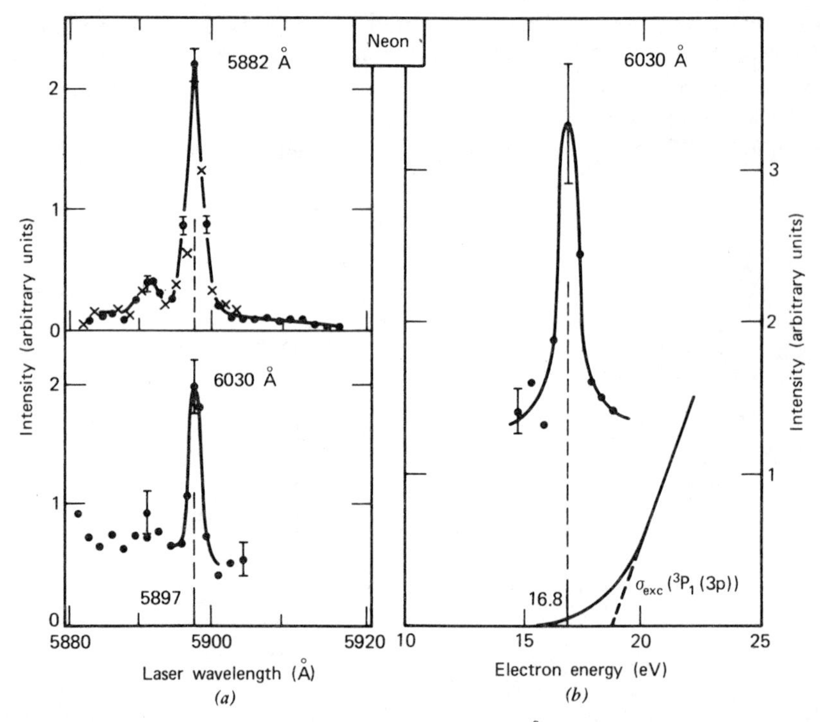

Fig. 4.37 Intensities of Ne *I* emission at 5882 and 6030 Å, induced by optical transitions from a Ne$^-$ level at 16.8 eV (see Fig. 4.35). (a) For a fixed electron energy of 16.8 ± 0.5 eV, as a function of dye laser wavelength. The largest peak occurs at 5897 Å; a smaller one lies at 5891 Å, i.e., at 2 meV separation. Full circles and crosses refer to separate data runs. (b) For a fixed laser wavelength of 5897 Å, as a function of electron energy. The $^3P_1(3p)$ excitation function (full curve), obtained as the total 6030 Å emission rate without gated counting, serves as a calibration of the energy scale and an indication of the energy width of the beam. (From Langendam, Gavrila, Kaandorp, and van der Wiel, 1976.)

left-hand diagram, the intensity of the Ne *I* emission at 5882 Å and 6030 Å is measured as a function of the dye laser wavelength at a fixed electron energy of 16.8 ± 0.5 eV. At a laser wavelength of 5897 Å, a narrow peak of 3 Å width is observed with a possible side-peak of lower intensity at 5891 Å. The interpretation is that in fact a transition has been induced between two Ne^- resonances which actually is a free-free transition of the scattering electron induced by the laser light. The observed width of 3 Å (corresponding to 1 meV) reflects the combined width of the two Ne^- resonances involved in the transition.

If this observation is real, this would be a powerful method for performing a spectroscopy on resonances in atoms and molecules with very high resolution (better than 1 meV). Some additional results have been reported at the ICPEAC in Paris (Gavrila, 1977), but it is certainly too early for a final judgement on this experiment.

2.4 Electron Scattering in Weak and Strong Radiation Fields

The final topic covers a class of processes that has become of increasing interest in recent years. Generally speaking, one is interested in collision processes in the presence of a radiation field. Here we concentrate on one particular process, the so-called free-free transitions. The processes of interest are:

$$
\begin{aligned}
e(E_i) + \mathrm{A} + \text{laser} &\rightarrow e(E_i + n \cdot \hbar\omega) + \mathrm{A} + \text{laser} &\text{(I)}\\
&\rightarrow e(E_i - b \cdot \hbar\omega) + \mathrm{A} + \text{laser} &\text{(II)}
\end{aligned}
$$

which correspond to free-free transitions of the scattering electron in the field of the atom A induced by the laser field. Process (I) represents induced absorption of n laser quanta, process (II) corresponds to induced emission of n quanta. These are important processes in astrophysics and hot laboratory plasmas, but they are also of considerable interest for the interaction of intense laser fields with matter. In the latter context, they are also of interest for molecules. The processes have been discussed in the literature (Gallagher, 1977), but until very recently there was no direct laboratory experiment for this type of process. This is now possible with the advanced technology of present electron spectrometers and high power lasers.

Work is in progress in three laboratories: (1) by Andrick and Langhans (1976, to be published) in Kaiserslautern, (2) by Weingartshofer, Holmes, Caudle, Clark, and Krüger (1977) in Antigonish, Nova Scotia/Canada, and (3) by Langendam, Gavrila, Kaandorp, and van der Wiel (1976) in Amsterdam. The Langendam–van der Wiel experiment which belongs in this category has been mentioned before, so the following discussion concentrates on some very recent results from the two other experiments.

From theoretical considerations in lowest-order approximation (Weingartshofer, Holmes, Caudle, Clarke, and Krüger, 1977; Krüger and Schulz, 1976) one obtains the following expression for the cross section of a free-free transition (one-quantum process $n = 1$):

$$\frac{d\sigma_{ff}^{(n=1)}}{d\Omega} = \frac{p_f}{p_i} \cdot \Gamma^2 \cdot \frac{d\sigma_{el}}{d\Omega} \tag{4.5}$$

where $d\sigma_{el}/d\Omega$ is the differential cross section for elastic scattering, p_f/p_i is a flux factor and Γ^2 is given by

$$\Gamma^2 = 4.86 \cdot 10^{-13} \cdot \lambda^4 \cdot F \cdot E_i \cdot \left(\frac{\boldsymbol{\epsilon} \cdot \mathbf{Q}}{2p_i}\right)^2. \tag{4.6}$$

In this relation, λ is the laser wavelength in μ, F is the flux density of the laser in W/cm^2, E_i is the energy of the incoming electron in eV, $\boldsymbol{\epsilon}$ is a unit vector characterizing the polarization of the laser light, and $\mathbf{Q} = \mathbf{p}_i - \mathbf{p}_f$ is the vector of momentum change for the scattering electron. The essential points are that the free-free cross section and the elastic cross section are expected to be proportional to each other and that the probability of a free-free process increases with the fourth power of λ. Furthermore, the factor $\boldsymbol{\epsilon} \cdot \mathbf{Q}$ is important for the choice of the experimental conditions. The experimental set-up used in the Andrick–Langhans experiment is shown in Fig. 4.38. In an electron scattering apparatus of essentially standard type an electron beam, an Ar atom beam and a CO_2 laser beam are crossed in the scattering center. Scattered electrons are detected in a plane perpendicular to the gas beam which contains the electron beam and the laser beam. The CO_2 laser was 90% polarized in this plane and the processes were observed in backward scattering (see the arrangement in Fig. 4.38). Note that under these conditions the factor $(\boldsymbol{\epsilon} \cdot \mathbf{Q}/2p_i)^2$ is close to 1. A CO_2 laser was chosen because of the λ^4 dependence of the free-free cross section. For a one-photon process the energy loss/gain is $\Delta E =$ 117 meV that can be conveniently resolved with the electron spectrometer. So the CO_2 laser represents the optimum compromise between the λ^4 increase of the cross section and the required energy resolution. Argon was chosen as the target gas, since the differential cross section for $e-$Ar scattering is fairly well known and gives a reasonable intensity in backward direction at energies around 10 eV.

A typical experimental result is shown in Fig. 4.39 (Andrick and Langhans, 1976, to be published). The energy distribution of the scattered

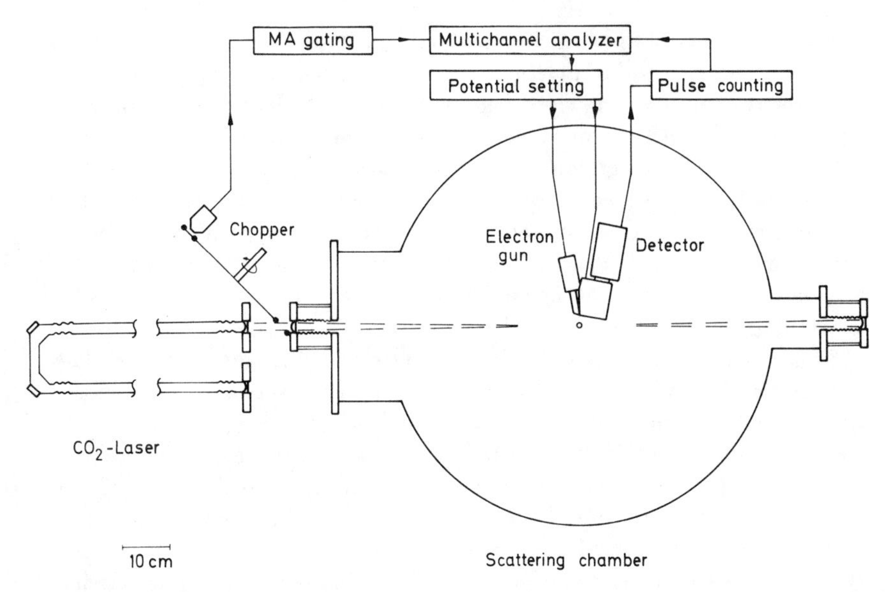

Fig. 4.38 Experimental setup of the Andrick and Langhans experiment. (From Andrick and Langhans, 1976.)

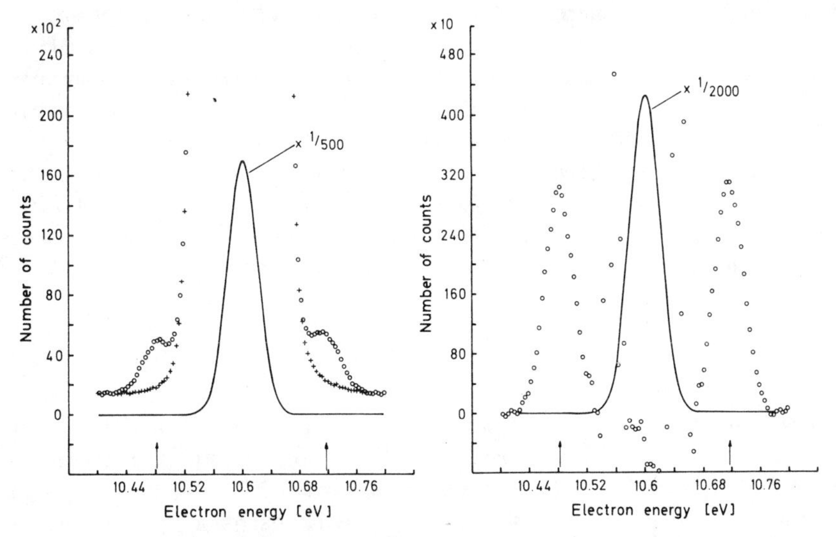

Fig. 4.39 Energy-loss spectrum for e-Ar scattering, circles with, crosses without laser beam. Number of counts per channel are plotted versus energy of the scattered electrons. Incident energy is 10.6 eV, scattering angle is 160°. The arrows indicate an energy gain or loss of 117 meV, the energy of one laser photon. The plot on the right-hand side shows the difference between both curves. The pure elastic peak is scaled down by the factors indicated. (From Andrick and Langhans, 1976.)

electrons is measured with and without the laser field. Small peaks are observed on both sides of the large elastic peak indicating one-photon free-free transitions. The right-hand side of Fig. 4.39 shows the difference signal together with the pure elastic peak which is scaled down by a factor of $\frac{1}{2000}$. For the comparison with theory according to (4.5) and (4.6), a careful evaluation is necessary mainly concerning the intersection volume of the three beams. One obtains reasonable agreement with theory (within error bars of about 25%) which is not too surprising, since with the 50 W CO_2 laser the experiment is performed under weak field conditions and everything can be calculated in first order.

This situation is changed in the experiment of Weingartshofer, Holmes, Caudle, Clarke, and Krüger (1977). The experimental set-up is very similar to the Andrick–Langhans experiment, but instead of using a 50 W CW-laser they use a pulsed CO_2 laser of 50 MW peak power. If the laser is focused into the scattering center, one obtains flux densities in the order of 10^9 W/cm^2. This corresponds to strong field conditions as we will discuss a little later.

The experimental result, otherwise under very similar conditions as in the Andrick–Langhans experiment, is shown in Fig. 4.40. The left-hand side of the figure shows the elastic peak without the laser, on the right-hand side we have the energy distribution of the scattered electrons with the laser on. There is clear evidence for multiphoton processes on both sides of the elastic peak. The free-free processes are not weak in intensity anymore. The total intensity (summed over all peaks) apparently remains constant, but a considerable redistribution of the energy is observed in the scattering process with the laser field.

The connection between both experiments can be written down in the following way. As a result from a theory called the semiclassical soft-photon approach (Weingartshofer, Holmes, Caudle, Clarke, and Krüger, 1977; Kroll and Watson, 1973) one obtains the formula

$$\frac{d\sigma_{ff}^{(n)}}{d\Omega} = \frac{p_f}{p_i} \cdot J_n^2(2\Gamma) \cdot \frac{d\sigma_{el}}{d\Omega} \tag{4.7}$$

for the free-free cross section where $J_n(2\Gamma)$ is the Bessel function of first kind and order n and Γ is given by equation (4.6). In both experiments we have $\lambda = 10.6\ \mu$, $E_i \approx 10$ eV, and $\epsilon \cdot Q/2p_i \approx 1$. The difference occurs in F, the flux density of the laser. In the Andrick–Langhans experiment we have $F \approx 2 \times 10^4$ W/cm^2 giving $\Gamma^2 \approx 10^{-3}$. Clearly, a power series of the Bessel function is possible which gives (4.5) for the $n = 1$ processes and negligible cross sections for multiphoton processes ($n > 1$). In the Weingartshofer experiment we have $F \approx 10^9$ W/cm^2 resulting in $\Gamma^2 \approx 50$. A power series is not meaningful and the oscillatory behaviour of the Bessel

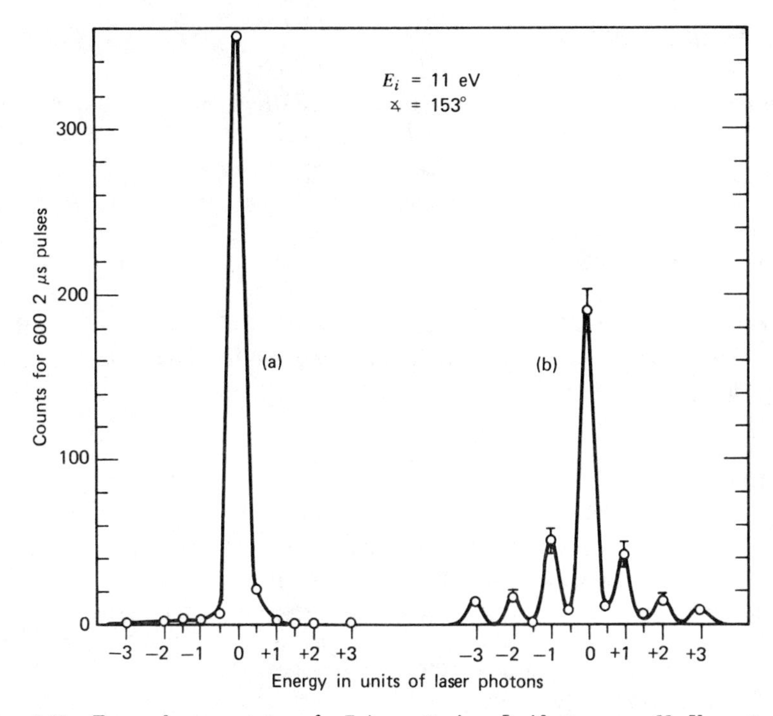

Fig. 4.40 Energy-loss spectrum of e^--Ar scattering. Incident energy 11 eV, scattering angle 153°. (a) Without laser field. The circles show the measured experimental points and the estimated outline of the process is drawn with a solid line, which was obtained by tracing out the elastic peak with a ratemeter and scaled to fit the maximum counts. (b) With laser field. The circles with error bars show the measured points and the estimated outline of the multiphoton (emission and absorption) processes are drawn in with solid lines obtained by scaling down the elastic peak as in (a). (From Weingartshofer, Holmes, Caudle, Clarke, and Krüger, 1977.)

function comes into play. These are the very first experiments of this kind. It would be very interesting to examine these ideas in improved and still better defined laboratory experiments.

There are further experiments of Andrick and Langhans which investigate the behavior of the free-free processes in the energy range of the $e-$Ar resonances around 11 eV. Very interesting effects are observed. The Langendam–van der Wiel experiment, mentioned previously, investigates free-free transitions between resonance states induced by laser light in the visible range. The whole field is in the very beginning and we expect a variety of new and interesting phenomena to be observed in future studies.

3 CONCLUSION

I hope I have shown you that this field, which was started by two important observations made by Schulz, is still a very exciting field and a playground for many beautiful laboratory experiments.

4 ACKNOWLEDGMENT

I am grateful to A. Gallagher, R. Hall, D. Andrick, L. Langhans, and K. Rohr for making material available before publication and for the permission to use it in this paper.

REFERENCES

Andrick, D. and L. Langhans, 1976, *J. Phys.*, **B9**, L459.

Andrick, D. and L. Langhans, to be published, *J. Phys.* **B**.

Andrick, D., L. Langhans, F. Linder, and G. Seng, 1975, Abstracts IX. ICPEAC, Seattle, University of Washington Press, p. 833.

Birtwistle, D. T. and A. Herzenberg, 1971, *J. Phys.* **B4**, 53.

Buckley, B. D., 1977, *J. Phys.* **B9**, L351.

Chang, E. S. and S. F. Wong, 1977, *Phys. Rev. Lett.*, **38**, 1327.

Crawford, O. H., 1967, *Proc. Phys. Soc.*, **91**, 279.

Crawford, O. H. and W. R. Garrett, 1977, *J. Chem. Phys.*, **66**, 4968, and references cited therein.

Dowell, J. T. and T. E. Sharp, 1968, *Phys. Rev.*, **167**, 124.

Dubé, L. and A. Herzenberg, 1977, *Phys. Rev. Lett.*, **38**, 820.

Gallagher, J. W., 1977, "Bibliography of Free-Free Transitions in Atoms and Molecules," JILA Information Center Report No. 16.

Gallagher, A. C. and G. York, 1974, *Rev. Sci. Instr.*, **45**, 662.

Gavrila, M., 1977, Invited Lectures of the X. ICPEAC, Paris, edited by G. Watel.

Hall, R., 1977, Invited Lectures of the X. ICPEAC, Paris, edited by G. Watel.

Hall, R. I., I. Cadez, C. Schermann, and M. Tronc, 1977, *Phys. Rev.*, **A15**, 599.

Herzenberg, A., 1968, *J. Phys.*, **B1**, 548.

Jordan, K. D. and J. J. Wendoloski, 1977, *Chem. Phys.*, **21**, 145.

Kroll, N. M. and K. M. Watson, 1973, *Phys. Rev.*, **A8**, 804.

Krüger, H., 1977, *Phys. Rev. Lett.*, **39**, 269.

Krüger, H. and M. Schulz, 1976, *J. Phys.* **B9**, 1899.

Kuyatt, C. E. and J. A. Simpson, 1967, *Rev. Sci. Instr.*, **38**, 103.

Langendam, P. J. K., M. Gavrila, J. P. J. Kaandorp, and M. J. van der Wiel, 1976, *J. Phys.*, **B9**, L453.

Linder, F., 1977, Invited Lectures of the X. ICPEAC, Paris, edited by G. Watel.

Rapp, D., T. E. Sharp, and D. D. Briglia, 1965, *Phys. Rev. Lett.*, **14**, 533.

Rohr, K., 1977, *J. Phys.*, **B10**, L399.

Rohr, K., to be published, *J. Phys.*

Rohr, K. and F. Linder, 1976, *J. Phys.*, **B9**, 2521.

Schulz, G. J., 1962, *Phys. Rev.*, **125**, 299.

Schulz, G. J., 1963, *Phys. Rev. Lett.*, **10**, 104.

Schulz, G. J., 1964, *Phys. Rev.*, **135**, A988.

Schulz, G. J., 1973a, *Rev. Mod. Phys.*, **45**, 378.

Schulz, G. J., 1973b, *Rev. Mod. Phys.*, **45**, 423.

Schulz, G. J., 1976, (reprinted as Chapter 1 of this volume) in *Principles of Laser Plasmas*, edited by G. Bekefi, Wiley-Interscience, New York.

Simpson, J. A., 1964, *Rev. Sci. Instr.*, **35**, 1698.

Tronc, M., F. Fiquet-Fayard, C. Schermann, and R. I. Hall, 1977, *J. Phys.*, **B10**, 305.

Turner, J. E., 1977, *Am. J. Phys.*, **45**, 758.

Turner, J. E., V. E. Anderson, and K. Fox, 1968, *Phys. Rev.*, **174**, 81.

Van Brunt, R. J., and A. C. Gallagher, 1977, Invited Lectures of the X. ICPEAC, Paris, edited by G. Watel.

Van Brunt, R. J., and A. C. Gallagher, 1977, Abstracts X. ICPEAC, Paris, Commissariat a l'Energie Atomique-Paris, p. 940.

Wallis, R. F., R. Herman, and H. W. Milnes, 1960, *J. Mol. Spectrosc.*, **4**, 51.

Weingartshofer, A., J. K. Holmes, G. Caudle, E. M. Clarke, and H. Krüger, 1977, *Phys. Rev. Lett.*, **39**, 269.

CHAPTER 5

The State of the Theory

NEAL LANE

Physics Department
Rice University
Houston, Texas

1 INTRODUCTION

In physics we strive toward the prediction of new phenomena not previously observed or new general physical laws not previously known. This is the excitement of the science. However, theoretical discoveries of this kind are usually not entirely accidental. It is often in the act of providing a better theoretical description of a process already studied experimentally that new discoveries are made. Even if nothing "earth shaking" comes out of a particular theoretical study, a good description of the phenomena still provides us with understanding at a more fundamental level—and that too is science. In this paper we (1) present a very brief overview of the theory with very little formal development, simply touching the important concepts, and (2) discuss a few selected examples where theoretical studies incorporating some of these ideas have been made. Since most progress has been made in elastic scattering and rotational and vibrational excitation, attention is restricted to these processes.

It is important to recognize at the outset that while all theorists have the same overall objective, they tend to go at it in rather different ways. The "*ab initio* theorist" attempts to solve the Schrödinger equation for scattering states of a system by methods which employ no adjustable parameters, effective potentials or "fudge factors" of any kind. However, some type of approximation is always required and the error is not easily estimated. Moreover, such precise studies are restricted to relatively simple systems at present. The "model theorist" considers more complicated systems or processes and makes use of model potentials, e.g., "exchange potentials" or "polarization potentials" to represent complicated effects. Unfortunately, the models involve "cutoff" and other parameters whose values are often chosen in a somewhat arbitrary manner. Finally, the "semi-empirical theorist" derives useful analytical relationships, e.g., angular distributions, effective-range formulas, which can be used to fit experimental data. Of course, such a division of theorists is arbitrary and certainly unrealistic. Actually, a single individual may work in any one of these modes depending on the problem at hand. The approaches complement one another and all are important to the advancement of the field. Here I concentrate on "near-*ab-initio*" and "model" studies.

2 CHARACTERISTICS OF ELECTRON-MOLECULE SCATTERING THEORY

A collision is inherently a time-dependent problem. However, in most experimental circumstances, the spread in the wave packet of the incident particle (electron here!) allows us to treat the process as time-independent, i.e., in a stationary-state approximation.

Thus, we seek continuum eigenfunctions $\psi(\mathbf{r}, x)$ of the total (electron plus molecule) Hamiltonian

$$H = -\tfrac{1}{2}\nabla_r^2 + V_{em} + H_m \tag{5.1}$$

where $\mathbf{r}$ denotes coordinates of the projectile electron and x denotes all internal coordinates, collectively, of the molecule; the terms H_m and V_{em} are the molecular Hamiltonian and the electron-molecule interaction, respectively. (Atomic units are assumed throughout and spin coordinates are implicit.) At large separations $r \to \infty$, we have $V_{em} \sim 0$ so that

$$\Psi(\mathbf{r}, x) \sim \Psi^{\mathrm{INC}}(\mathbf{r}, x) + \Psi^{\mathrm{SCATT}}(\mathbf{r}, x) \tag{5.2}$$

where the first term represents an incident plane wave with the molecule in its initial state $\psi_0(x)$ and the second term is a superposition of scattered-wave contributions of the form

$$\Psi^{\mathrm{SCATT}}(\mathbf{r}, x) = \frac{1}{r}\sum_n \exp(ik_n r) f_{n0}(\hat{r})\psi_n(x) \tag{5.3}$$

where k_n^2 is the final kinetic energy (in rydbergs) of the inelastically scattered electron leaving the molecule in state $\psi_n(x)$, and $f_{n0}(\hat{r})$ is the scattering amplitude for this process $0 \to n$. The index n represents internal quantum numbers ν (electronic), v (vibrational), and j and m_j (rotational) for the simple case of a diatomic target. The differential cross section for the process $0 \to n$ is simply expressed in terms of the scattering amplitude by

$$\left(\frac{d\sigma}{d\Omega}\right)_{0\to n} = \frac{k_n}{k_0}|f_{n0}(\hat{r})|^2. \tag{5.4}$$

The most natural approach would seem to be a so-called "coupled-states" or "close-coupling" expansion in terms of the complete set of target states $\{\psi_n(x)\}$ of the molecule. This takes the form

$$\Psi(\mathbf{r}, x) = \mathcal{A}\sum_n F_n(\mathbf{r})\psi_n(x) \tag{5.5}$$

where, assuming that the proper spin couplings have been included, $\mathcal{A}$ is the familiar antisymmetrization operator. The coefficients $F_n(\mathbf{r})$ in such an expansion play the role of continuum scattering wave functions. However, they are coupled to one another via the "infamous" coupled partial integro-differential equations

$$(\nabla_r^2 + k_n^2)F_n(\mathbf{r}) = 2 \sum_{n'} [V_{nn'}(\mathbf{r}) + K_{nn'}(\mathbf{r})]F_{n'}(\mathbf{r}) \tag{5.6}$$

where the "direct" matrix elements are defined by

$$V_{nn'}(\mathbf{r}) \equiv \int \psi_n(x)^* V_{em}(\mathbf{r}, x)\psi_{n'}(x)\, dx \tag{5.7}$$

and the "exchange" matrix elements are operators that "pull" the scattering function $F_{n'}(\mathbf{r})$ under the integral and "eject" a bound molecular orbital in its place

$$K_{nn'}(\mathbf{r})F_{n'}(\mathbf{r}) = \left\{\int_0^\infty X_{nn'}(\mathbf{r}, \mathbf{r}')F_{n'}(\mathbf{r}')d^3r'\right\}\phi_c(\mathbf{r}) + \cdots, \tag{5.8}$$

where $\phi_c(\mathbf{r})$ is a bound molecular orbital. The function $X_{nn'}(\mathbf{r}, \mathbf{r}')$ is well behaved; its precise form is not important here. Since $\phi_c(\mathbf{r})$ is a bound orbital it falls off exponentially at large r. This is true of all the terms in (5.8) and one describes exchange as a short-range interaction. The direct matrix elements, in contrast, include important long-range contributions.

There are four sources of rather serious complication in attempting to solve (5.6):

1. Rotational and vibrational response of the nuclei.
2. Nonspherical character of V_{em} $(\mathbf{r}, x)$.
3. Nonlocal exchange interaction.
4. Correlation via excited electronic states.

The first two on the list are problems that we do not have in the case of electron-atom scattering. The large number of vibrational and rotational states that are energetically accessible at energies well below the first electronic threshold forces us to retain a correspondingly large number of terms in (5.5) and (5.6) even in the "simple" case of the static-exchange approximation, where all excited electronic states are ignored. (We shall see later that under most circumstances of interest, application of the Born–Oppenheimer approximation to the electron-plus-molecule system simplifies the nuclear response problem immensely.) The anisotropy problem, i.e., nonspherical character of the electron-molecule interaction, remains a formidable obstacle. The ordinary partial-waves expansion of the functions $F_n(\mathbf{r})$ in terms of spherical harmonics $Y_{lm}(\hat{r})$, which reduces the coupled partial differential equations to coupled ordinary differential equations, does not work very well here. The difficulty is that the scattering functions $F_n(\mathbf{r})$ exhibit strong dependence on θ as well as r, especially for values of the electron coordinates $\mathbf{r} = (r, \theta, \phi)$ near to one of the nuclei. The expansion in terms of $Y_{lm}(\hat{r})$ is therefore slowly convergent, i.e., we need a large number of partial waves! In principle, two-center coordinates will help in the case of diatomic molecules; in practice the situation is still not so clear. For polyatomic molecules, two-center coordinates are not helpful. Recently, several so-called "L^2 Methods" (e.g., R-matrix, T-operator, pseudo-bound-state or low-l spoiling methods) have been developed to deal with this problem by employing techniques common to molecular structure theory. Some applications are discussed later on. Finally, the problems of exchange and correlation, while similar to the case of electron-atom scattering, are further compounded by complications (1) and (2) discussed above. In spite of the complexity of the problem, real progress has been made.

2.1 Nuclear Impulse (Adiabatic Nuclei) Approximation

The nuclear response problem can essentially be eliminated by employing the nuclear impulse approximation (Chang and Temkin, 1970), valid when the effective collision time τ_c is much less than the periods τ_R or τ_v associated with nuclear rotation or vibration, respectively. Since typically $\tau_R \sim 10^{-12}$ sec and $\tau_v \sim 10^{-14}$ sec, the approximation may work well for electron energies as low as $\sim 10^{-1}$ eV or even lower in the case of rotation. The difficulty in determining precisely how well the approximation will work in a given case rests on the assignment of an appropriate collision time. Generally, the impulse approximation tends to fail:

1. Near thresholds.
2. Near resonances.
3. For polar molecules.

The nuclear impulse approximation is simply based on the Born–Oppenheimer approximation, where the electron-molecule scattering problem is first solved for the nuclei fixed, say with separation R and orientation $\hat{R}$ in the case of a diatomic molecule (Schneider, 1976). The fixed-nuclei scattering wave function may be obtained from an (electronic) close-coupling expansion

$$\Psi^{(e)}(\mathbf{r}, \mathbf{r}_i; \mathbf{R}) = \mathscr{A} \sum_{\nu} F_\nu(\mathbf{r}; \mathbf{R}) \psi_\nu^{(e)}(\mathbf{r}_i; R) \tag{5.9}$$

where $\mathbf{r}_i$ stands collectively for bound electrons in the target, the $\psi_\nu^{(e)}$ are electronic wave functions of the target molecule, and the index ν runs over all such electronic states. The functions $F_\nu(\mathbf{r};\mathbf{R})$ depend only parametrically on the nuclear coordinates $\mathbf{R}$ that are fixed for each calculation; these functions satisfy a set of coupled equations similar in form to (5.6), but involving only electronic states. Since $\mathbf{R}$ is fixed, it is convenient to carry out the full calculation of the $F_\nu(\mathbf{r}; \mathbf{R})$ in the body frame of the molecule. If only a single electronic state is retained in (5.9), the coupled equations (5.6) reduce to a single equation; this is the static exchange approximation. The full scattering wave function, including the nuclear degrees of freedom, may then be written in the Born–Oppenheimer approximation

$$\Psi(\mathbf{r}, \mathbf{r}_i, \mathbf{R}) \cong \Psi^{(e)}(\mathbf{r}, \mathbf{r}_i; \mathbf{R})\chi^{(n)}(\mathbf{R}) \tag{5.10}$$

where $\chi^{(n)}(\mathbf{R})$ is simply the unperturbed nuclear (rotation and vibration) wave function corresponding to the initial state. The final target state is simply projected out of the asymptotic form for Ψ in (5.10). One obtains for the approximate scattering amplitude for a transition $0 \rightarrow n$, i.e., $(\nu_0 v_0 j_0 m_{j0} \rightarrow \nu v j m_j)$ the simple expression (Chase, 1956)

$$f_{n0}(\mathbf{r}) \cong \langle vjm_j | f^{(e)}_{\nu\nu_0}(\hat{\mathbf{r}}, \mathbf{R}) | vjm_{j0} \rangle, \tag{5.11}$$

where $f^{(e)}_{\nu_0}$ is the fixed-nuclei (electronic) scattering amplitude derived from $\Psi^{(e)}$ of (5.9). Thus, for the case of rotational and vibrational excitation within the ground electronic state of a molecule, one need only calculate the "elastic" fixed-nuclei scattering amplitude at a number of R values; a simple rotational transformation to the lab frame introduces the $\hat{R}$ dependence. The integration over $\mathbf{R}$ is then straightforward.

The impulse approximation as stated in (5.11) fails near thresholds since no account is taken of energy transfer between the electron and molecule during a transition. Near threshold, the initial- and final-state continuum wave functions are very different and the correct scattering amplitude reflects this (Shugard and Hazi, 1975). A treatment incorporating a frame transformation (Chang and Fano, 1972) at an intermediate electron-molecule separation eliminates the difficulty near threshold and still takes advantage of the simplicity of a body frame, fixed-nuclei, treatment at least when the electron is close to the molecule. At larger separations, where transformation to the lab frame is necessary, all coupling is weaker, and complicated interactions such as exchange are no longer present. The frame-transformation procedure also eliminates, in principle, the problem with polar molecules since the energy splittings of the rotational states are properly included when the electron is far from the molecule. The unrealistic divergence in the forward direction of the fixed-nuclei scattering amplitude for electron-polar-molecule scattering does not enter. Unfortunately, there have been few actual applications of the frame-transformation method, and cases which have been studied exhibit unexpected complications.

2.2 Resonant Scattering

The problem in dealing with resonances (especially narrow resonances) is simply that the electron can stay around long enough to alter the potential energy of the nuclei. In the "adiabatic limit" of long-lived resonances, where the lifetime $\tau_r \gg \tau_v$, the nuclei tend to establish vibrational states of the negative molecular ion ($e + O_2$ is a prototype), and the $\chi^{(n)}(\mathbf{R})$ may differ substantially from the unperturbed nuclear wave functions. Even for intermediate cases where $\tau_r \cong \tau_v$, the nuclei are not well described by unperturbed nuclear wave functions and the impulse approximation is inaccurate ($e + N_2$ is a prototype).

Herzenberg and Mandl (1962) demonstrated that a resonant-scattering formalism of the Siegert (or Kapur and Peierls) form was appropriate under such circumstances. The Born–Oppenheimer approximation is still applied, where one writes the total wave function in product form

$$\Psi(\mathbf{r}, \mathbf{r}_i; \mathbf{R}) = \Psi_\alpha^{(e)}(\mathbf{r}, \mathbf{r}_i; \mathbf{R})\eta^{(n)}(\mathbf{R}) \tag{5.12}$$

but where the electronic wave function $\Psi_\alpha^{(e)}$ is not a continuum eigenfunction as in (5.9) and (5.10), but rather satisfies the complex eigenvalue problem

$$[H_e - W_\alpha(R)]\Psi_\alpha^{(e)}(\mathbf{r}, \mathbf{r}_i; \mathbf{R}) = 0 \tag{5.13}$$

where the eigenvalues

$$W_\alpha(R) = E_\alpha(R) - \frac{i}{2}\Gamma_\alpha(R), \qquad \alpha = 1, 2, \ldots \tag{5.14}$$

represent complex potential energy curves on which the nuclei move. The nuclear wave functions $\eta^{(n)}(\mathbf{R})$ satisfy equations that include both "growth" and "decay" terms. In a time-independent formalism, these terms allow for the time-dependent processes of capture of the electron into the negative-molecular-ion state α, response of the nuclei to the new field, and decay of the resonance state back to stationary states of the target molecule. Recent applications of this method are discussed below.

An alternative approach to resonant scattering is the elegant Feshbach Projection Operator formalism, in which the space is partitioned by writing

$$\Psi(\mathbf{r}, x) = (P + Q)\Psi(\mathbf{r}, x) \tag{5.15}$$

where $Q\Psi$ is usually taken to satisfy a real eigenvalue problem of the form

$$(QHQ - \epsilon_i)Q\Psi = 0 \tag{5.16}$$

the roots of which are approximate positions of resonances in the scattering problem. The function $P\Psi$ describes the scattering, and via coupling terms QHP, picks up the resonant contributions as well as the background scattering. Formal developments valuable to the understanding of electron-molecule scattering have been made (Chen, 1969), but complete treatments are still awaited.

A technique that provides a good estimate of the real part of the resonant potential energy curves is the Stabilization Method of Taylor (1970). The beauty of the method is that it employs standard molecular-structure variational procedures applied to the negative molecular ion. It is noted that certain energies $E_r(R)$ exhibit a stability as more variational freedom is allowed, provided the basic set is sufficiently flexible to permit representation of diffuse nonresonant scattering states with energies below as well as above the "resonance energy curve" $E_r(R)$. Successful application of the method is somewhat of an art; still, it has contributed greatly to our understanding of electron-molecule resonant scattering, in particular the complicated process of dissociative attachment. In principle, the method may be extended to obtain the "width," hence lifetime $\Gamma(R)$, and background scattering information as well. We now understand the nature of the scattering processes under a wide range of conditions, and recognize the limitations of the various theoretical approaches. What we have yet to agree on is the most convenient way to handle electron-molecule scattering that permits a smooth transition between nonresonant and resonant conditions in a single problem. We have not yet put together a "single theory" that bridges the gap in an optimum manner.

3 SELECTED RECENT APPLICATIONS OF THE THEORY

In the limited time available here, we focus on "near-*ab-initio*" calculations and "model studies" applied to elastic scattering and rotational and vibrational excitation at low energies, $E \lesssim 10$ eV. At these energies the static electron-molecule interaction and the electron-exchange and polarization (long-range correlation) interactions are all important. Recent developments include: (1) alternative mathematical approaches, the so-called "L^2 methods" that employ bound-state, molecular-structure techniques to extract scattering information; these include R-Matrix Theory, T-Matrix Expansion, Generalized Pseudopotential (Pseudo-Bound-State) Method, and (2) approximation methods aimed at simplifying the treatment of the electron-exchange interactions, e.g., Free-Electron-Gas (FEG),

Semi-Classical-Exchange (SCE), Core-Orthogonalization-Exchange (COE), to name a few. The polarization interaction is included either via a polarized orbital treatment *á la* Temkin or by introducing a local polarization potential with some semiempirical cut-off parameter. So far, the latter approach has been necessary for all targets more complicated than H_2.

A brief comment on the general nature of these interactions might be useful here. For electron collisions with closed-shell, ground-state (electronic) molecules at energies below the first electronic threshold, both the exchange and polarization interactions are "attractive" in nature. The "static exchange" term from (5.6) and (5.8) becomes

$$K(\mathbf{r})F(\mathbf{r}) = \left\{ \int_0^\infty X(\mathbf{r}, \mathbf{r}')F(\mathbf{r}')d^3r' \right\} \phi_c(\mathbf{r}) + \cdots \quad (5.17)$$

in the fixed-nuclei case, and when only the ground electronic state is included. The effect of such a term is to place a "Fermi hole" around each target-orbital electron of the same spin as the projectile, thus reducing the screening of the nuclei by these electrons and resulting in a more attractive potential energy. The polarization interaction, on the other hand, arises from an adiabatic redistribution of molecular charge (induced dipole) in response to the electric field of the projectile electron, at least when the electron is still far away from the molecule. As the electron approaches the nuclei, its "local speed" increases and the adiabatic picture breaks down. One recipe for dealing with this difficulty is simply to ignore any redistribution of bound-electron charge resulting from the projectile electron's penetration inside the bound orbitals. While this has been successfully applied to $e - H_2$ scattering (Hara, 1969a, Henry and Lane, 1969), it is complicated and does not carry over conveniently to more complicated target molecules. A common semiempirical procedure is simply to add on a local "polarization potential" of the form

$$V_{\mathrm{POL}}(\mathbf{r}) - \left\{ \frac{\alpha_0}{2r^4} + \frac{\alpha_2}{2r^4} P_2(\cos\theta) \right\} C(r) \quad (5.18)$$

where

$$C(r) = 1 - \exp\left[-\left(\frac{r}{r_c} \right)^6 \right] \quad (5.19)$$

and α_0 and α_2 are the spherical and "nonspherical" polarizabilities; the "cut-off" parameter r_c forces V_{POL} to zero for all $r \ll r_c$. Several specific examples are discussed below. Of course, if all electronic states were retained in a close-coupling expansion such as that in (5.5) or the corresponding fixed-nuclei expansion in (5.9), these polarization effects would be automatically included as "virtual excitations" and the further addition of a polarization potential would be incorrect. However, in practice, convergence is slow as it is in e-atom scattering so that other methods are required.

3.1 H_2: Elastic Scattering

Although H_2 is a rather simple example since it has only two bound electrons and is not a very strong scatterer, nevertheless, it has received a considerable amount of attention and much of what we have learned in these studies carries over to other cases. In Fig. 5.1 (adapted from Massey, 1969) calculated total cross sections for $e-H_2$ scattering in the static approximation (II_a and III_a) and the static-exchange approximation (II_b and III_b) are compared with the experimental measurements. Several points are worth noting. These are fixed-nuclei calculations with the nuclear separation frozen at $R=1.4a_0$ (equilibrium) and the z axis taken along the body axis $\hat{R}$. A partial-waves expansion of $F(\mathbf{r})$ shows that only a few partial waves are important and in fact the $s\sigma$ partial wave, i.e., the

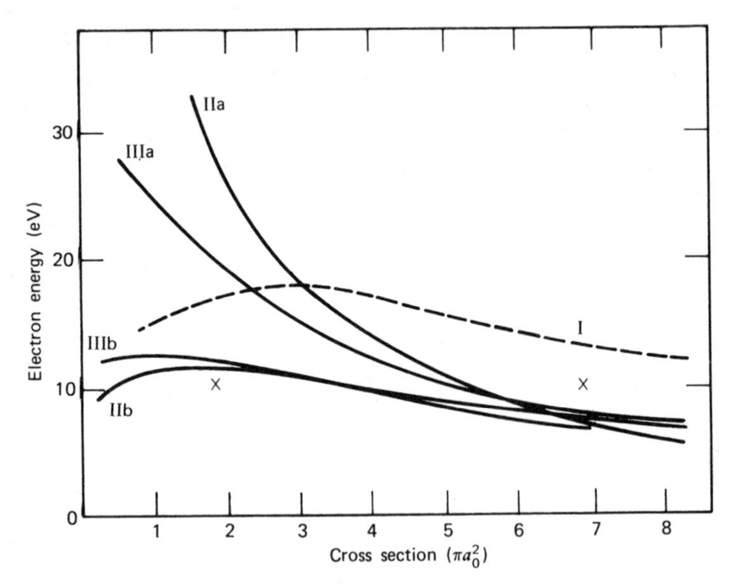

Fig. 5.1 Comparison of observed and calculated cross sections for collisions of electrons with H_2 molecules. I observed by Golden, Bandel, and Salerno; × calculated by Nagahara; IIa calculated by Massey and Ridley—exchange ignored; IIb calculated by Massey and Ridley—exchange included; IIIa calculated by Hara—exchange ignored; IIIb calculated by Hara—exchange included. (From Massey, 1969.)

partial wave corresponding to $m = 0$ and even parity that is primarily s-wave asymptotically, dominates the low energy scattering. The effect of the exchange interaction (included approximately in these calculations) is to lower the cross section at low energies by increasing the σ_g eigenphase shift closer to π.

In Fig. 5.2 (adapted from Golden et al., 1971) somewhat more precise static-exchange calculations by Tully and Berry (1969), Hara (1969a) ("no polarization") and Henry and Lane (1969) ("close coupling (no polarization)") are compared with calculations including both exchange and polarization as reported by Hara ("including polarization") and Henry and Lane ("close coupling"). All of these calculations fix the internuclear separation at the equilibrium value $R = 1.4a_0$. The calculations of Henry and Lane allow explicitly for nuclear rotation in the close-coupling calculation. The other calculations take R to be fixed and average the cross section over nuclear orientation. Under conditions where the nuclear impulse approximation is valid, this should give the same result as

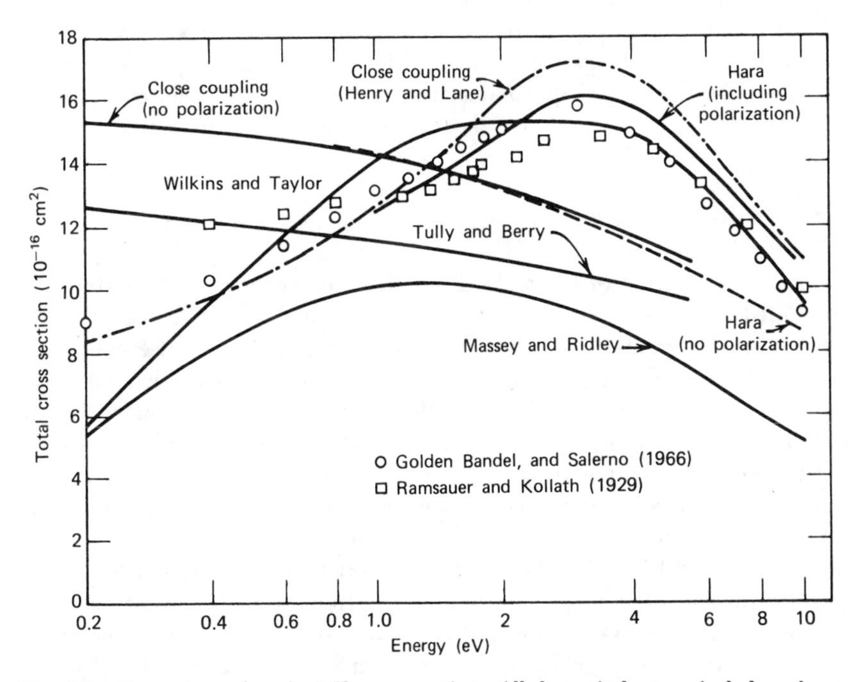

Fig. 5.2 Comparison of total e^--H_2 cross sections. All theoretical curves include exchange in one form or another. Theoretical calculations plotted: Henry and Lane (1969) close coupling, with polarization (dot-dash), and without (solid); Hara (1969a) two center, with polarization (solid), and without (dash); Wilkins and Taylor (1967); Massey and Ridley (1956); Tulley and Berry (1969). (From Golden et al., 1971.)

summing the elastic and rotational-excitation cross sections over all possible final states. Chang and Temkin (1970) first showed, by careful analysis of the model close-coupling calculations of Lane and Geltman (1967), that the rotational impulse approximation works well for H_2. The effect of polarization is to further reduce the low-energy cross section while increasing the cross section, in particular the $p\sigma$ contribution at higher energies. The experimental data of Ramsauer and Kollath (1929) and Golden, Bandel, and Salerno (1966) are given for comparison. The calculated differential cross sections, not shown here, are also in very good agreement with experiment when both exchange and polarization effects are included, but not when either is ignored.

Accurate static-exchange calculations based on the R-matrix (Schneider, 1975) and T-matrix-expansion (Resigno, McCurdy, and McKoy, 1974, 1975) methods have also been carried out for $e-H_2$ scattering. In the latter case, the method has been successfully extended to include polarization effects via an optical potential calculated by means of a perturbation expansion (Kaldor and Klonover, 1977).

Not shown here are recent results obtained by Baille and Darewych (1977) using a number of different approximate exchange potentials. They conclude, in partial agreement with Riley and Truhlar (1975) that the semiclassical exchange approximation is superior to the electron-gas potentials, but that none of these approximations are reliable below $\approx$6 eV.

3.2 H_2: Rotational Excitation

Rotational excitation of the molecule is also a low-energy process since the rotational spacing is so small (for H_2, $B = 60\ \text{cm}^{-1} = 0.0074$ eV). The mechanism for rotational excitation is however quite different from elastic scattering. A very long time ago Massey (1932) pointed out that for polar molecules, rotational excitation near threshold would be dominated by the long-range e-dipole interaction; he also noted that the Born approximation should work well there (except possibly for very large dipole moments) since the electron wave function would only be weakly perturbed by the interaction. Gerjuoy and Stein (1955) carried this argument over to nonpolar molecules and Dalgarno and Moffett (1963) suggested one could include the anisotropic polarization interaction in a similar manner. At higher energies, it was expected that short-range interactions (including exchange) would play an important role. Ardill and Davison (1968) first showed in their distorted waves calculation that exchange was important to rotational excitation at higher energies.

In Fig. 5.3 (adapted from Takayanagi and Itikawa, 1970), theoretical and experimental rotational excitation cross sections for $j = 0 \rightarrow 2$ in

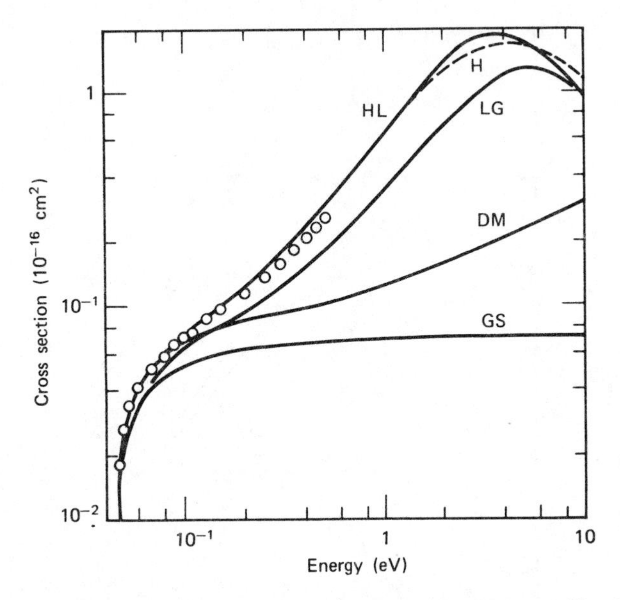

Fig. 5.3 The experimental and theoretical excitation cross sections σ ($J=0\rightarrow 2$) of H_2. HL: Henry and Lane (1969). H: Hara (1969b) LG: Lane and Geltman (1967). DM: Dalgarno and Moffett (1963). GS: Gerjuoy and Stein (1955). Circles are experimental data obtained by Crompton et al. (1969). (From Takayanagi and Itikawa, 1970.)

para-H_2 are compared. Near threshold the Born–Quadrupole (GS) and Born–Quadrupole–Polarization (DM) work well. At higher energies, distortion of the partial waves and short-range coupling become important as illustrated by the model close-coupling studies of Lane and Geltman (1967). The more precise (Body-Frame; Rotational Impulse) calculations of Hara (1969b) and (lab-frame) calculations of Henry and Lane (1969), both including polarization and exchange effects, are in good agreement with experiment, at least for energies $E \lesssim 0.6$ eV where comparison is possible.

At higher energies the situation is less clear. In Fig. 5.4 (adapted from Takayanagi and Itikawa, 1970) cross sections for $j=1\rightarrow 3$ in ortho-H_2 are compared. The latest measurement, viz. that of Linder (1969) is expected to be the best currently available. At energies above $\cong$1 eV, the results of Hara (1969b) "H_1" and Henry and Lane (1969) "HL" are clearly different, and neither agrees well with the experimental data. The other curves "H_2"–"H_4" of Hara (1969b) illustrate the importance of exchange and polarization effects in the rotational-excitation process.

It should be emphasized that in all of these theoretical treatments, the internuclear separation was held fixed at $R \simeq 1.4a_0$. The extent to which

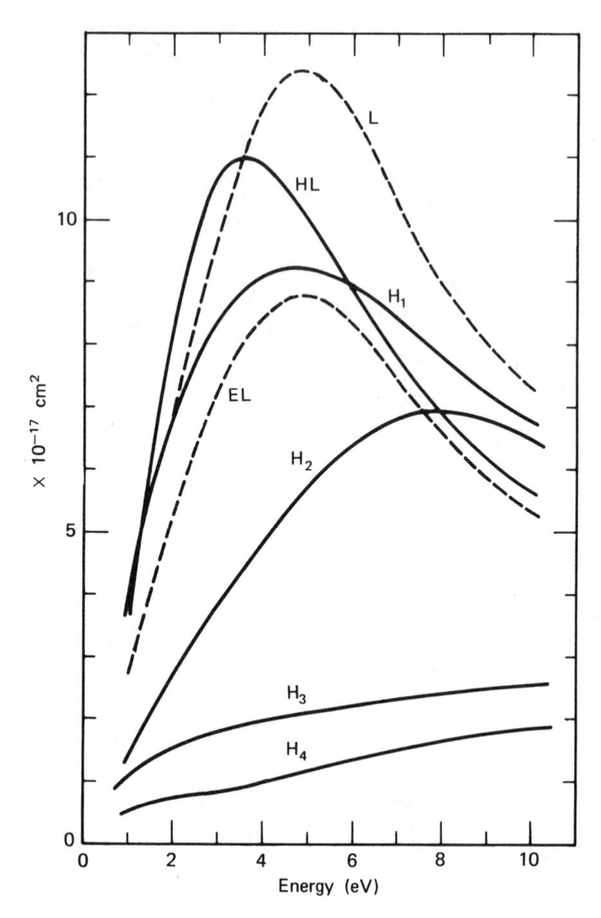

Fig. 5.4 The experimental and theoretical excitation cross sections σ ($J = 1 \rightarrow 3$) of H_2. HL: Henry and Lane (1969). H_1–H_4: Hara (1969b), calculations with full interactions, without polarization, without exchange, and with static interaction only, respectively. Experimental curves are EL (Ehrhardt and Linder, 1968), and L (Linder, 1969). (From Takayanagi and Itikawa, 1970.)

allowance for vibrational motion during the collision or even averaging the cross sections over R might change the results is not known. In fact vibrational excitation in H_2 has received very little theoretical attention.

Electron scattering from H_2 is relatively "weak" compared to that from a number of other molecules. A more characteristic molecule is N_2 where strong anisotropy in the interaction is evident and where the overall interaction is strong enough to support a shape resonance. The experimental total cross section measured by Golden (1966) is shown in Fig. 5.5

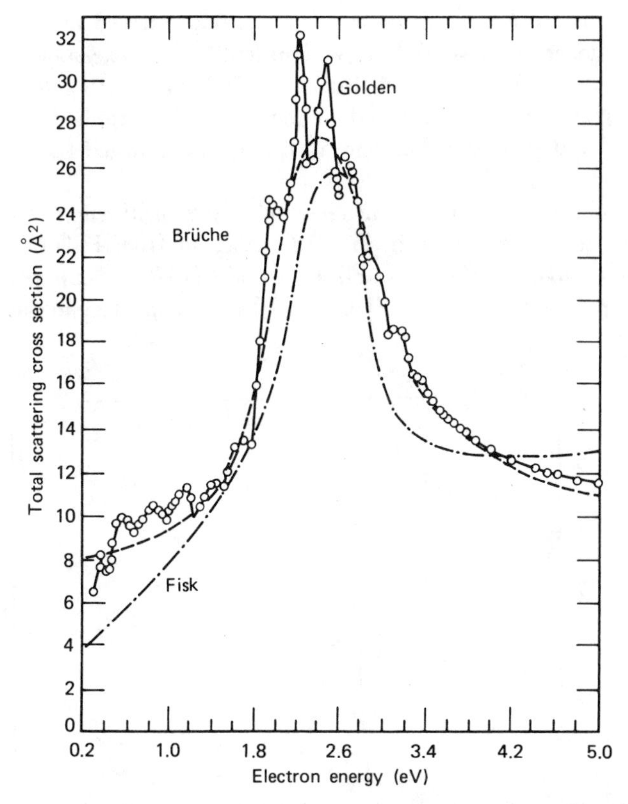

Fig. 5.5 Total scattering cross sections in square angstroms for e^{-}-N_2, from Golden (1966), Brüche (1927), and Fisk (1936). (From Golden et al., 1971.)

(adapted from Golden et al., 1971) along with early model-potential calculations. The low-energy structure is apparently not real. However, the features superimposed on the peak are due to vibrational response during the collision.

3.3 N_2: Elastic Scattering

The treatment of N_2 is considerably more complex than that of H_2 primarily due to the stronger and more anisotropic e-N_2 interaction. This results in a slowly convergent single-center expansion of the wave function. While a two-center approach such as Crees and Moores (1975) should be better, the resulting equations are hard to handle, and at present we still do not have an unambiguous comparison.

An approximate (nonconverged) static-exchange calculation was first carried out by Burke and Sinfailam (1970), but suffered from convergence problems. Recently, several "more-complete" static-exchange calculations have been done. Buckley and Burke (1977) solved the static-exchange coupled equations, Morrison and Schneider (1977) applied the R-matrix method and Fliflet, Levin, Ma, and McKoy (1977) applied their T-operator expansion method.

All authors agree "semi-quantitatively". Just to illustrate the nature of this cross section, we show in Fig. 5.6 (adapted from Fliflet et al., 1977) the static-exchange result of Fliflet et al. (1977) compared with the nonconverged Burke and Sinfailam (1970) result and the measurements of

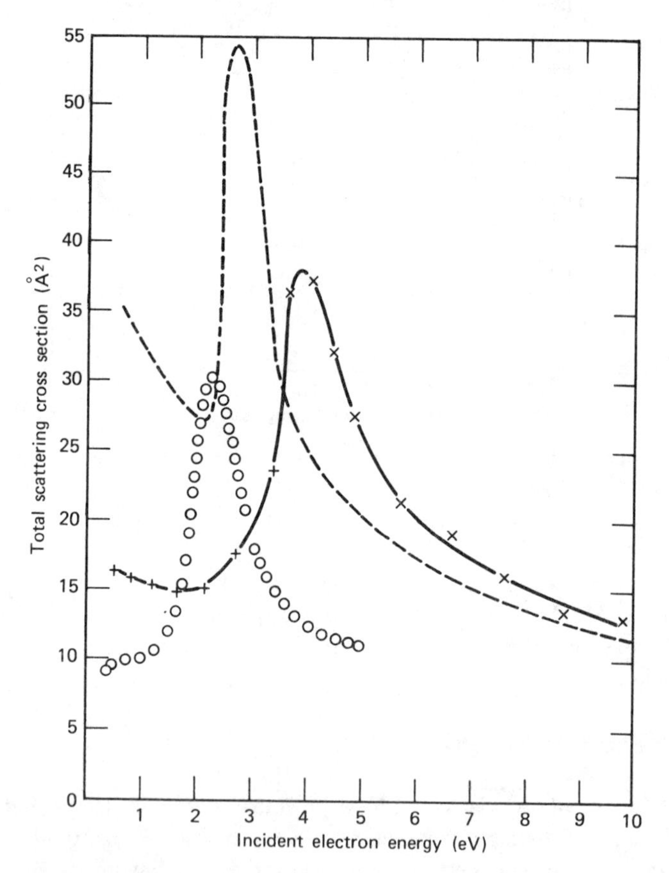

Fig. 5.6 Total cross sections for e-N_2 scattering. Theory: solid curve and ×, + from T-matrix static exchange calculations of Ffiflet et al. (1978); dashed curve from unconverged static-exchange calculations of Burke and Sinfailam (1970). Experiment: open circles from Golden (1966); observed vibrational structure not shown here.

Golden (1966) (the fine structure is not shown). Other recent static-exchange results are similar.

Buckley and Burke (1977) included an effective polarization potential in their calculation to account partially for this important correlation effect. The long-range r^{-4} form was smoothly cut off at a radius ($r_c = 2.308a_0$) chosen to "tune" the position of the resonance to the experimental resonant energy. Figure 5.7 (adapted from Morrison and Collins, 1977) compares their approximate polarization result "*BB*" with Golden's data "*G*."

An alternative to including exchange via explicit antisymmetrization of the wave function is to use a model exchange potential, a pseudopotential,

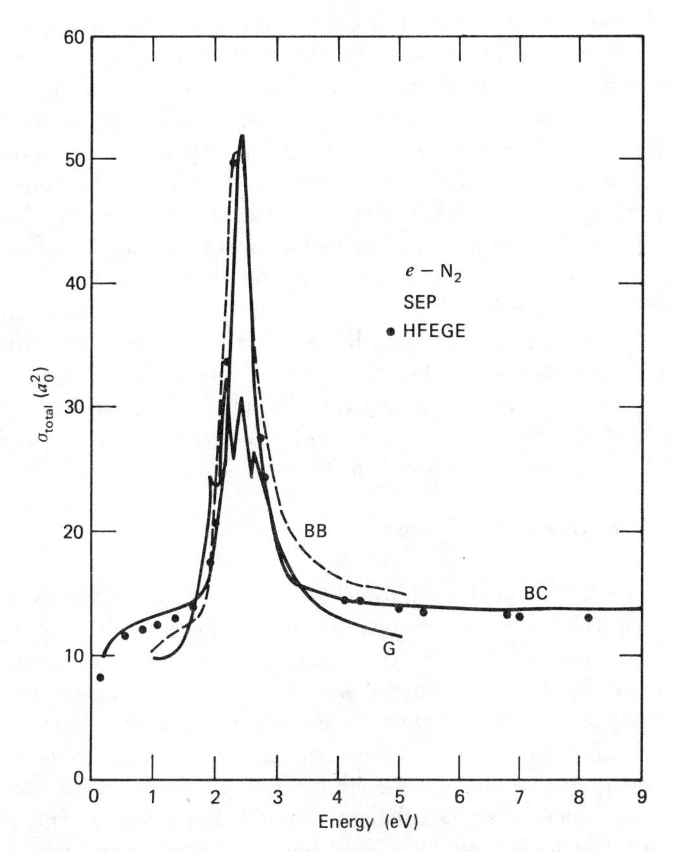

Fig. 5.7 Total cross sections for e-N_2 scattering. Theory: static-exchange-polarization calculations; dashed curve BB from Buckley and Burke (1977); solid curve BC from Burke and Chandra (1972); closed circles from Morrison and Collins (1978). Experiment: solid curve G from Golden (1966).

or the equivalent. Burke and Chandra (1972) adopted a pseudopotential approach that was based on the notion that the most important exchange effect is to force orthogonality of the scattering orbital to bound fully-occupied orbitals of like symmetry. Polarization was included via the "parametrized" model potential cut-off to "tune" the resonance to its observed energy. Their total cross section "*BC*" is also included in Fig. 5.7.

Morrison and Collins (1977) included exchange via a local (Hara free electron gas exchange) potential similar to Slater's $\rho^{1/3}$ potential but modified in the manner suggested by Hara (1967) in an early study of H_2. They do not force orthogonality as in the Burke and Chandra approach. Results are also shown "*HFEGE*" in the figure. Polarization is treated in the same manner, however.

These three calculations, i.e., "*BB*," "*BC*," and "*HFEGE*" have two important features in common. They all include polarization via a local polarization potential cut-off at some radius chosen to "tune" the position of the calculated resonance to occur near the center of the measured resonance structure. Second, the internuclear separation is fixed at the equilibrium value throughout the calculation. Thus, the observed structure, which arises from the nuclear response to the N_2-resonant-complex, is not present in the calculations. In order to recover this structure, nuclear vibrational motion must be included. The structure is, of course, present in rotational-excitation as well as elastic cross sections.

Chandra and Temkin (1976) extended the model pseudopotential approach of Burke and Chandra by explicitly treating the vibrational close-coupling problem for the π_g symmetry where the resonance occurs. In the other symmetries they apply the vibrational impulse approximation. This "hybrid theory" results in a total cross section shown in Fig. 5.8 (adapted from Chandra and Temkin, 1976).

3.4 N_2: Vibrational Excitation

In order for vibrational excitation to occur, it is necessary that the coherence of the initial vibrational state be somehow destroyed. In the impulse limit, the colliding electron "impulsively" alters the phase of the wave function in a manner that depends on the internuclear separation. This "phase interruption" results in a mixing of vibrational states; thus, if energetically allowed, a vibrational transition may occur. In the resonant, compound-molecular-ion limit where a long-lived electronic state is formed, the nuclei respond adiabatically to a new potential energy curve, viz. that of the molecular ion, resulting in a mixture of possible final vibrational states of the target. The case of e-N_2 scattering around $E = 2$ eV lies between these limiting cases.

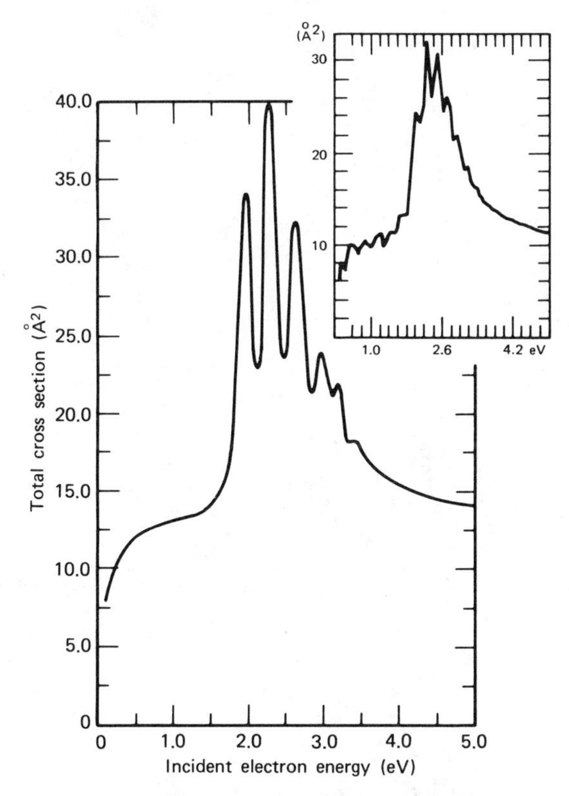

Fig. 5.8 Total scattering cross section in the full hybrid theory. Inset, experimental results of Golden. The experimental structure below 1.8 eV is not found and is considered to be spurious. Golden (private communication) has suggested that his low-energy structure may be due to resonant superelastic collisions with vibrationally excited N_2 present in his target gas. Our calculation assumes all N_2 is initially in the ground vibrational state. (From Chandra and Temkin, 1976.)

The early theoretical studies of Herzenberg and Mandl (1962), Chen (1964), and Hasted and Awan (1969), based on the idea of a compound N_2^- state obtained good qualitative agreement with the unusual vibrational excitation cross sections first measured by Schulz (1962). The overall structure was fairly well reproduced by the theories, although the precise resonance formation mechanism was in some dispute. Figure 5.9 (adapted from Birtwistle and Herzenberg, 1971) shows the comparison.

Birtwistle and Herzenberg (1971) demonstrated that a number of systematic features in these e-N_2 vibrational excitation cross sections could be better described by allowing for the variation of the resonance

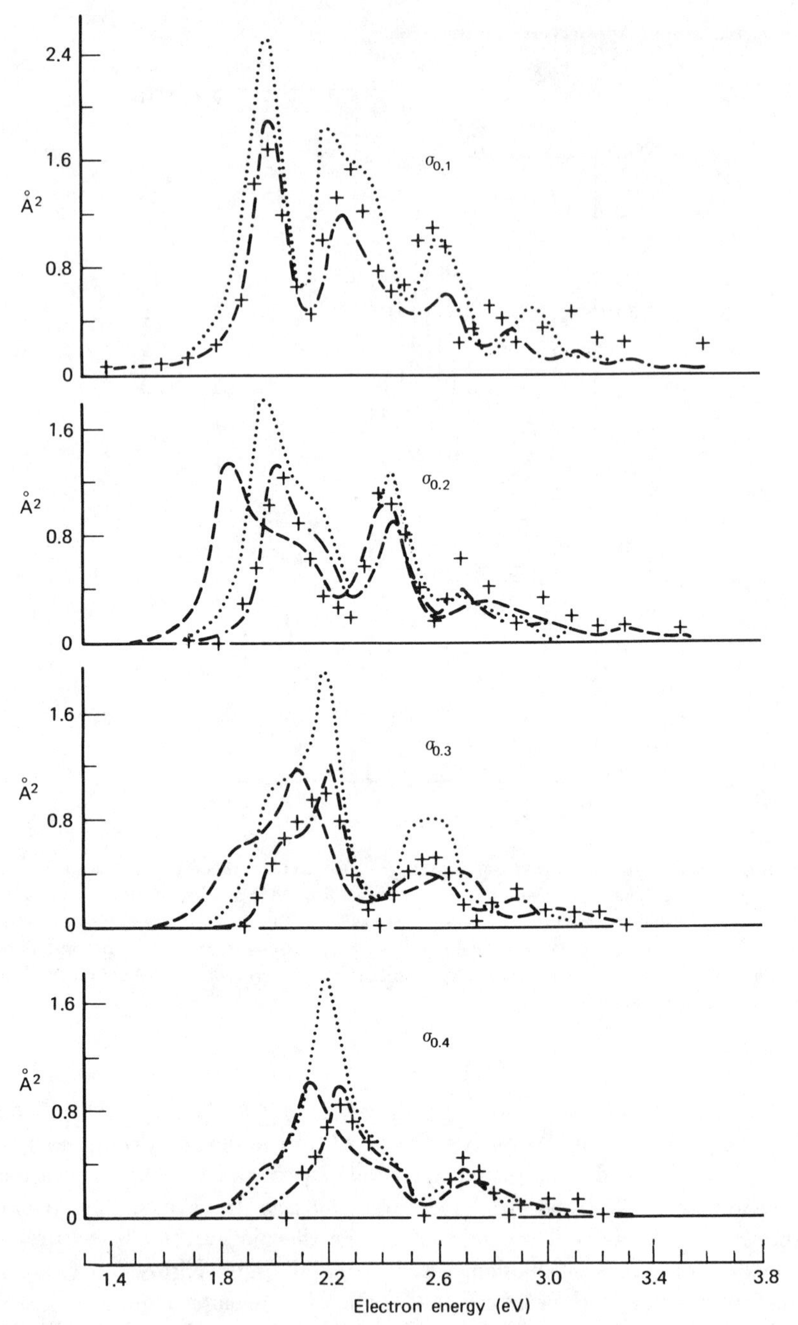

Fig. 5.9 Cross sections for resonant e-N_2 scattering. σ_{0v} is the cross section for the excitation of the vth vibrational state, starting from the ground, that is the 0th state. Crosses: experiment (Schulz, 1962). The curves give the results of various published calculations: broken curve, Herzenberg and Mandl (1962); chain curve, Chen (1964); dotted curve, Hasted and Awan (1969); crosses and broken curve normalized to chain curve.

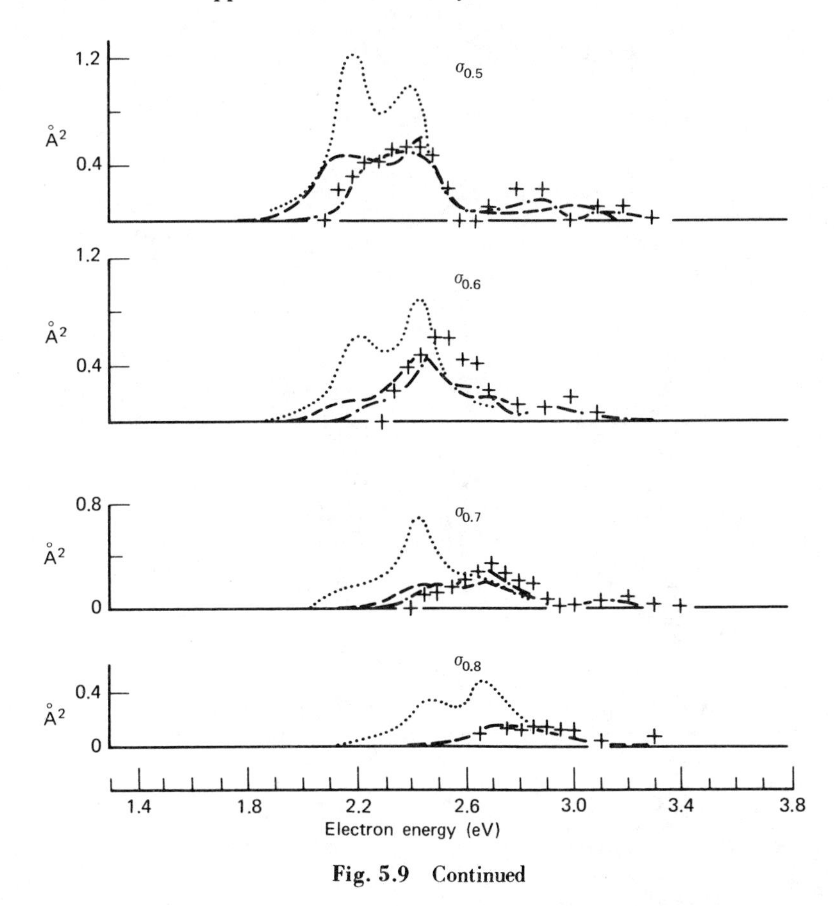

Fig. 5.9 Continued

lifetime with internuclear separation. In Fig. 5.10, their so-called "boomerang model" is illustrated. It turns out that the width $\Gamma(R)$ is so large, hence the lifetime $\tau(R) \propto \Gamma(R)^{-1}$ so small, that the nuclei (represented by wave packets in the figure) have time for roughly only one reflection at the outer repulsive wall of the N_2-potential. The interference of outgoing and reflected nuclear wave packets successfully accounts for the complicated observed structure, as illustrated in Fig. 5.11 (adapted from Birtwistle and Herzenberg, 1971; see section 3.1 of this article for details concerning the parameters used in the calculation). While the theory is semiempirical in the sense that parameters in the width $\Gamma(R)$ and N_2-potential $V_{-}(R)$ are chosen to best fit the data, good agreement for all the transitions is not necessarily guaranteed, and suggests the mechanism is correctly described.

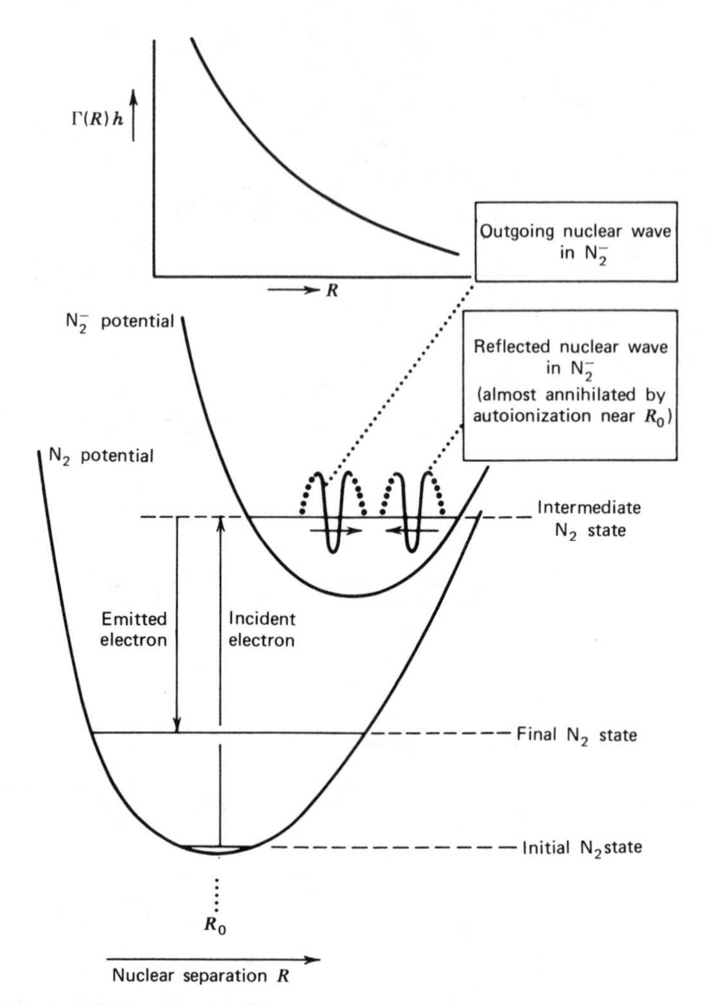

Fig. 5.10 The "boomerang" model of the nuclear wave function in the N_2-ion, which suggests the calculation with variable Γ reported here. This schematic model is discussed in an earlier paper (Herzenberg, 1968). It is based on the assumption that the magnitude and R dependence of $\Gamma(R)$ are such that only a single outgoing and a single reflected wave matter. (From Birtwistle and Herzenberg, 1971.)

An ambitious calculation of vibrational excitation in N_2 has been carried out by Chandra and Temkin (1976) (we discussed elastic cross sections from this work earlier). The authors approximate exchange by an explicit orthogonalization to core orbitals (as did Burke and Chandra) and polarization by an effective R-dependent model potential. In their hybrid

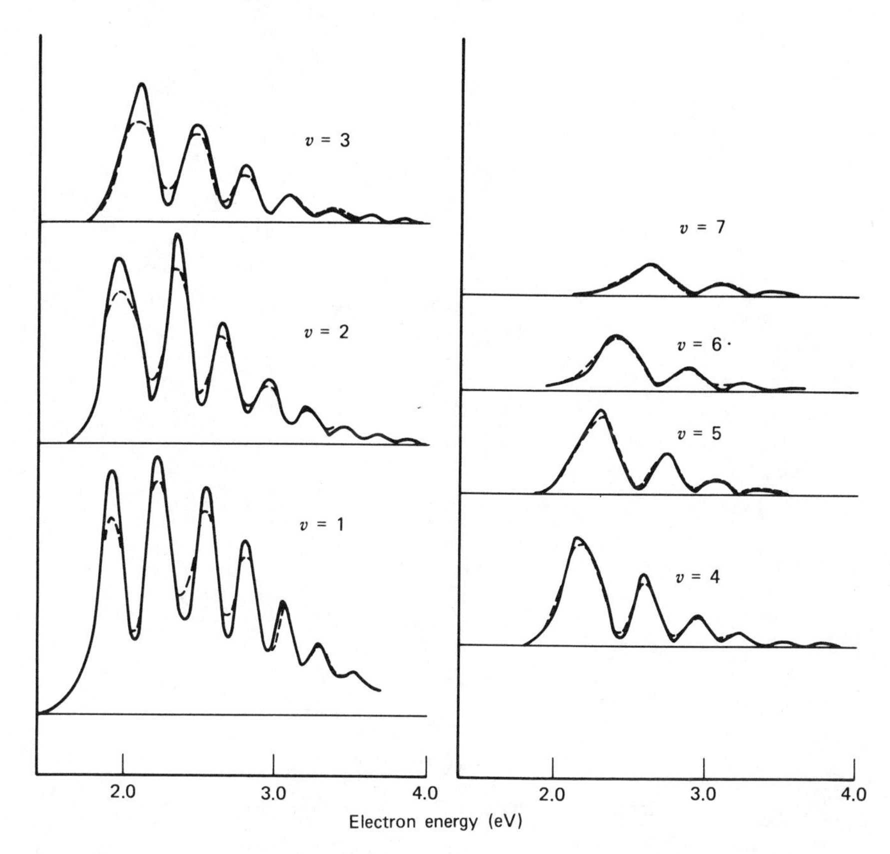

Fig. 5.11 Comparison of experimental cross sections (Ehrhardt and Willman, 1967, solid curve) with the cross sections calculated for the parameters stated in Sec. 3.1 (broken curve). Arbitrary units; only relative cross sections are significant. (From Birtwistle and Herzenberg, 1971.)

theory Chandra and Temkin solve coupled vibrational-state equations for the resonant π_g symmetry; for the other symmetries they employ the vibrational impulse approximation. Results are shown in Fig. 5.12 (adapted from Chandra and Temkin, 1976) for several vibrational excitation cross sections; the experimental results of Schulz (1964) (increased by factor of 2) are shown for comparison. There is still uncertainty in the proper experimental normalization; the shapes of the cross-sectional curves are of more interest here. Qualitative agreement is obtained for the lower transitions, but deteriorates rapidly as one goes to larger final-state quantum numbers v'. According to the authors, the problem is very likely their inability to converge the coupled-channels calculations in terms of

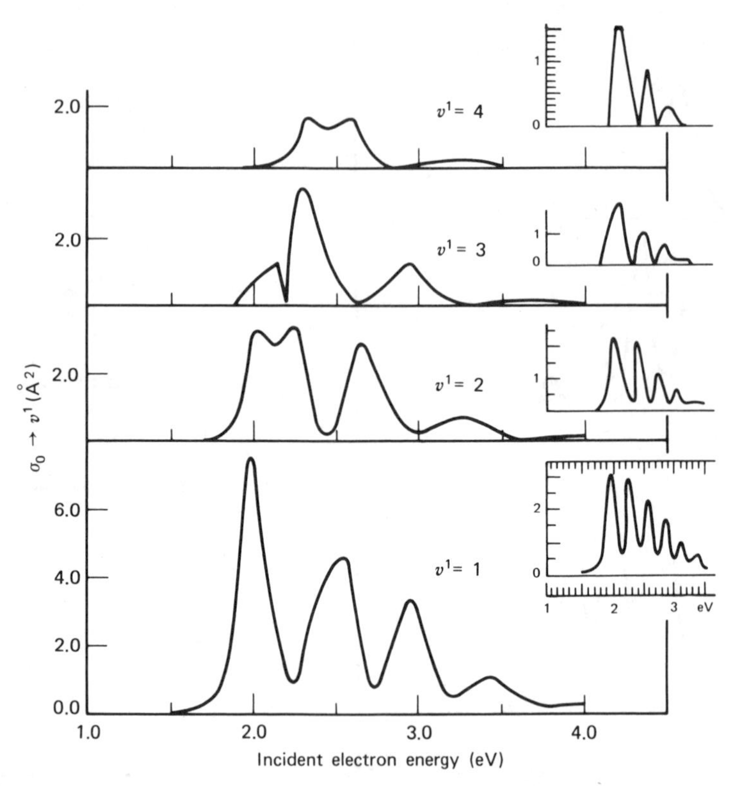

Fig. 5.12 Total vibrational excitation curves. (From Chandra and Temkin, 1976.)

the number of partial waves l or vibrational states v included in the expansion.

Clearly, it is very difficult to carry out a converged close-coupling calculation (converged in l and v) even for a model representation of the e-N_2 interaction, where exchange and polarization effects are treated only approximately. Moreover, the positions and widths of the resonances and their variation with final state v are very subtle interference phenomena, requiring high accuracy in any calculation that does not explicitly parametrize these features.

In any case, we do have a fairly good understanding of the resonance region, and we are in a good position to estimate the difficulty of improving the calculations by degrees. (A frame-transformation calculation where the inner region is treated in the fixed-R approximation, and vibrational close-coupling used in the outer region may be a useful next step; but, it will still be difficult.)

3.5 F_2: Elastic Scattering

It should be briefly mentioned that two recent applications of new *ab initio* methods to e-F_2 scattering. Rescigno, Bender, McCurdy, and McKoy (1976) have carried out an approximate static-exchange calculation using a discrete basis set approach. Assuming that different partial waves are effectively decoupled at large electron separations, the scattering phase shifts are obtained directly from an ordinary Hamiltonian matrix diagonalization using a large basis with diffuse functions. They obtain a relatively flat cross section at low energies with a broad Σ_u^+ shape resonance at about 4 eV.

Schneider and Hay (1976) independently applied R-matrix theory to this problem, also in the static-exchange approximation. They obtained similar results. Beyond this they showed that a "reasonable" degree of correlation could cause the Σ_u^+ resonance to become bound resulting in a smooth cross section. Recent dissociative attachment experiments favor the latter.

3.6 CO_2: Elastic Scattering

In attempting to describe electron scattering processes involving even relatively simple polyatomic molecules such as CO_2, one is faced with complications due to additional degrees of nuclear freedom, increased density of nuclear states, and stronger anisotropy in the electron-molecule interaction potential. Electron-CO_2 scattering has been the subject of a fair amount of experimental investigation and some theoretical attention as well. It is a good case for further theoretical study.

The low-energy elastic cross section has been known to be unusually large since the early measurements of Ramsauer and Kollath (1927). Since the equilibrium geometry of ground-state CO_2 is linear, it has no dipole moment. Speculation about its strong scattering properties have pointed to its large quadrupole moment ($\sim -3.8ea_0^2$) or perhaps a "transient dipole" effect connected with zero-point bending of the molecule.

Recently Morrison, Lane, and Collins (1977) carried out an approximate "static-exchange-with polarization" (SEP) calculation with fixed nuclei, very similar to the N_2 work of Morrison and Collins (1977) already discussed. Exchange is included as a local free-electron-gas potential modified in the manner of Hara (there are no adjustable parameters in the exchange potential). A polarization potential is cut off at a radius ($r_c = 2.59a_0$) chosen to "tune" the characteristic resonance to 3.8 eV as was first done by Burke and Chandra for e-N_2 scattering.

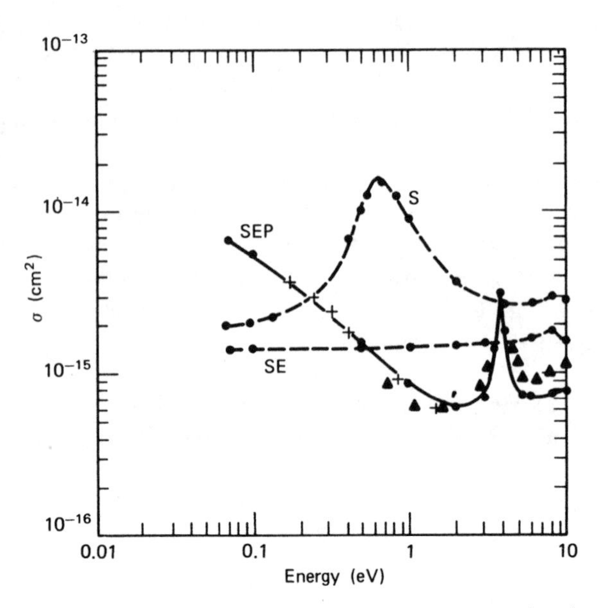

Fig. 5.13 Converged total integrated cross sections (including Σ_g, Σ_u, Π_g, and Π_u, symmetries) for the e-CO_2 scattering in the static (S), static exchange (SE), and static-exchange-polarization (SEP) models. Experimental data: +.

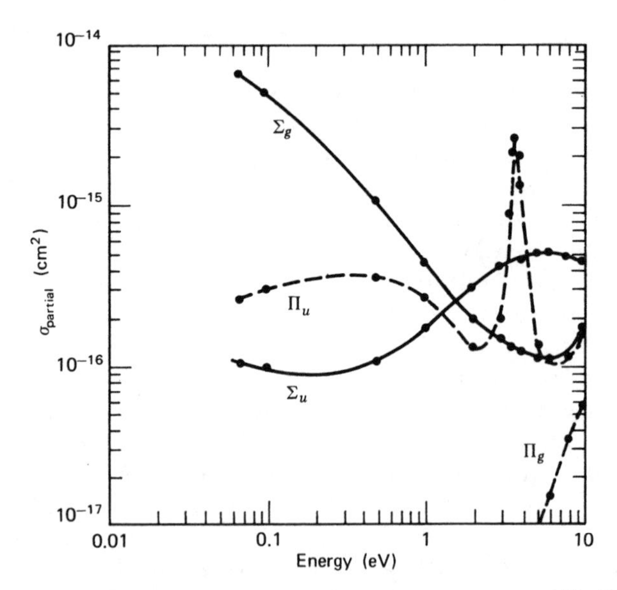

Fig. 5.14 Converged integrated cross sections for e-CO_2 scattering in Σ_g, Σ_u, Π_g, and Π_u symmetries in the SEP model using the HFEGE potential and a polarization cutoff radius $r_c = 2.59a$. The Π_u resonance energy is 3.8 eV.

The calculated static-exchange-polarization (SEP) calculations of the total e-CO_2 cross section is compared in Fig. 5.13 with static-exchange (SE) and static (S) calculations and with early measurements of Ramsauer and Kollath (1927) (+) and Brüche (1927) (▲). The π_u resonance, barely evident in the SE results at $E \cong 8$ eV, is "tuned" to 3.8 eV by including the polarization. The calculated resonance is too narrow and too high, reflecting failure of the fixed-nuclei approximation for energies in the resonance region. Apparently, CO_2 is a "boomerang" case similar to N_2, and should exhibit a broad "fine structure."

Relative contributions of the important symmetries are illustrated in Fig. 5.14. The low-energy scattering is dominated by Σ_g which at large e-CO_2 separations is primarily s-wave in character. The scattering length of the dominant $k \to 0$ eigenphase shift is negative and suggests a modest "virtual-state" enhancement of the Σ_g cross section. (This enhancement may give rise to resonances in vibrational-excitation just above threshold, as have been observed in a number of molecules.) The Σ_u and π_u contributions are dominated by the electron-quadrupole interaction at low energies giving way to shorter-range interactions resulting in a "potential resonance" for the π_u case. Calculated momentum-transfer cross sections

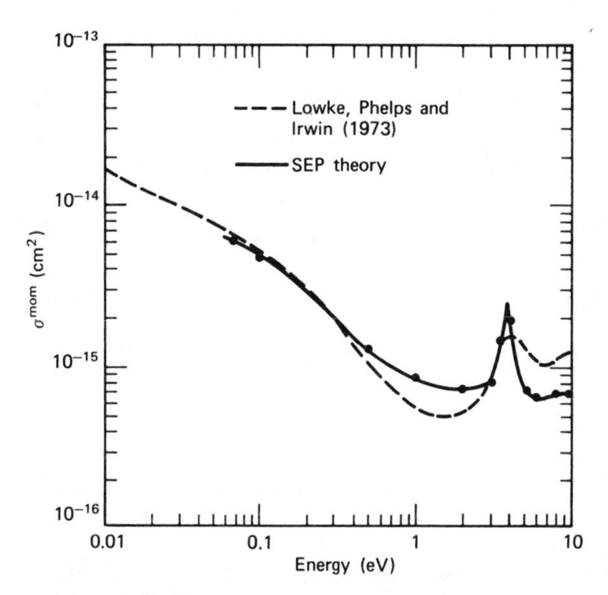

Fig. 5.15 Converged momentum-transfer cross sections for the e-CO_2 scattering in the SEP model using the HFEGE potential with $r_c = 2.59a_0$ compared with experimental data from swarm experiments.

are compared with the data of Lowke, Phelps, and Irwin (1973) in Fig. 5.15.

Rotational-excitation cross sections of Morrison and Lane (1977) were obtained by applying the rotational-impulse (adiabatic-nuclear-rotation) approximation. In Fig. 5.16, calculated $j = 0 \rightarrow 2$ cross sections are com-

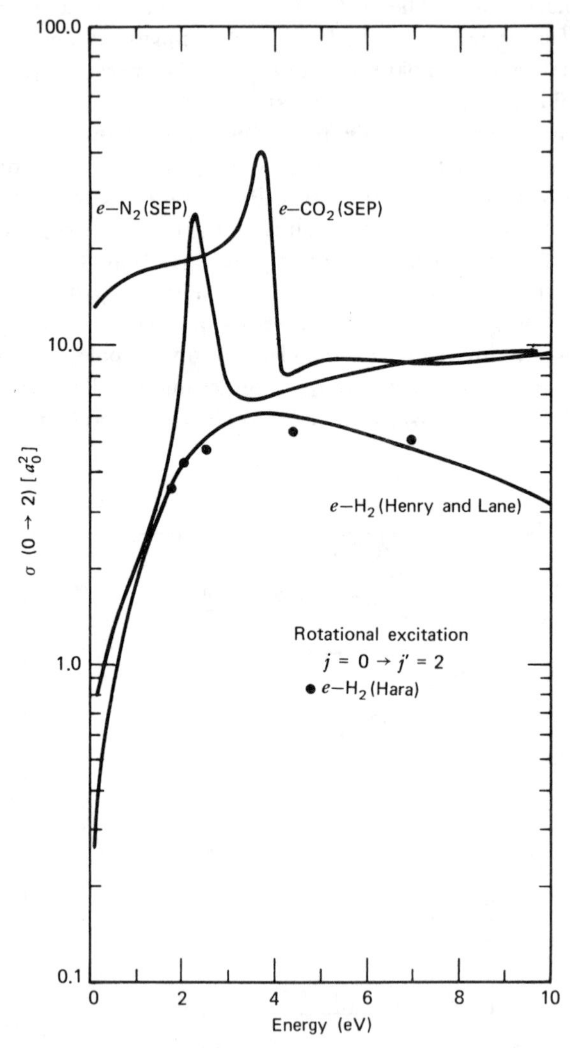

Fig. 5.16 Comparison of theoretical rotational-excitation cross sections for $j = 0 \rightarrow 2$ transitions in e-H_2, N_2, and CO_2 scattering. Curve e-CO_2 SEP from Morrison and Lane (1977); curve e-N_2 SEP from Morrison and Collins (1978); curve e-H_2 from Henry and Lane (1969); closed circles from Hara (1969b).

pared for CO_2, N_2, and H_2. The e-N_2 results are those of Morrison and Collins (1977) and the e-H_2 cross sections are those of Hara (1969b) and Henry and Lane (1969). With proper allowance for nuclear vibration, both the e-N_2 and e-CO_2 cross sections should exhibit a fine structure. At very low energies, rotational-excitation is dominated by the e-quadrupole interaction. The Born approximation, valid and very near threshold, predicts a Q^2 dependence, where the ratios of the squares of the quadrupole moments Q^2 for $H_2:N_2:CO_2$ are roughly (0.25):(1.0):(16.0). As the energy is increased from threshold, short-range interactions quickly become important.

3.7 CO_2: Vibrational Excitation

No detailed *ab initio* calculations have been carried out for vibrational excitation of CO_2. However, experimentally observed cross sections have been convincingly interpreted in two ways: (1) the "boomerang model" of Birtwistle and Herzenberg (1971) applied to CO_2 by Cadez (1977) and (2) an alternative parameterization scheme of Domeke and Cederbaum (1977) based on a novel partitioning technique. The agreement with experiment is very similar to that obtained by Birtwistle and Herzenberg for N_2.

To the extent that the scattering is dominated by the resonant CO_2^- state (π_u at linear geometry) certain features can be understood by considering the CO_2^- potential energy curves. Claydon, Segal and Taylor (1970) first calculated these curves applying the semiempirical INDO method. More precise calculations were carried out by Krauss and Neumann (1972) and Pacansky, Wahlgren and Bagus (1975). These structure studies have been a vital key to the qualitative understanding of vibrational excitation of different modes in CO_2.

At lower energies, i.e., well below the resonance region (3.8 eV) our theoretical understanding is much less satisfactory. Born approximations have been carried out by Itikawa (1971) using approximate long-range interactions. One might expect Born to be good very near threshold at least for the optically allowed bending or asymmetric stretch transitions since a long-range, weak coupling is involved (much like rotational excitation near threshold). However, there is evidence that low-energy e-CO_2 scattering exhibits a "virtual-state enhancement." This could invalidate the Born results and give rise to sharp resonant behavior very near the thresholds. Model calculations are currently underway at Rice University to help resolve this question. Sharp resonances have been observed in a number of polar molecules and the virtual state mechanism has been invoked by Dubé and Herzenberg (1977) to explain these results.

3.8 Polar Molecules: Elastic and Rotational Excitation

Polar molecules are best considered in a class by themselves. The long-range r^{-2} electron-dipole interaction potential dominates the total cross section at low energies since important contributions arise from a large number of partial waves and distant scattering is important.

At low (thermal) energies the total cross sections are dominated by rotational excitation rather than elastic scattering and this rotational excitation is mainly due to the long-range dipole interaction. The differential cross section is strongly forward peaked. For polar molecules with small to medium dipole moments (say $\mu \leq 1ea_0$), the Born-dipole approximation is adequate (Massey, 1932) at low energies since it properly treats the long-range scattering (i.e., higher partial waves) and therefore yields accurate forward-scattering cross sections, which dominate the integral cross section. The case of HCl is illustrated in Fig. 5.17.

The momentum-transfer cross section σ_m, however, includes a weighting factor $1 - \cos\theta$ which deemphasizes forward scattering and may, therefore, be more sensitive to short-range interactions including exchange. Early

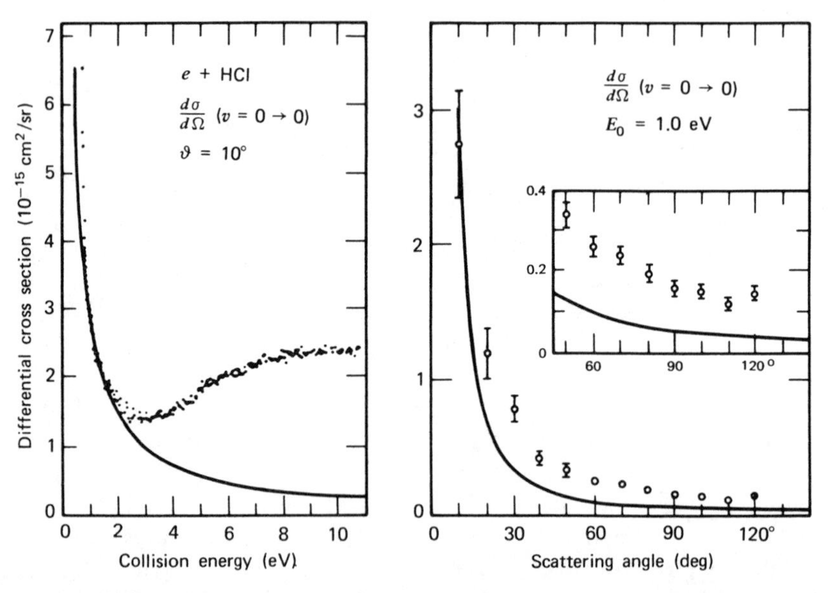

Fig. 5.17 Differential cross sections for vibrationally elastic scattering from HCl. Left: Energy dependence at $\vartheta = 10°$. Right: Angular dependence at $E = 1$ eV. Experimental results are compared with the Born approximation (full line). (From Rohr and Linder, 1977; Linder, 1977.)

interpretation of average momentum-transfer cross sections (thermal energies) suggested that molecules possessing dipole moments $D \cong$ 1.62 debye might have unusually large values of σ_m, in which case it seemed that even σ_m was controlled by the long-range interaction. ($D_c \cong$ 1.62 debye is the so-called "critical dipole moment," the minimum value of D such that a fixed dipole can bind an electron.) Subsequent analysis of the rotating molecule with and without short-range interactions, argues convincingly that no unusual behavior of σ_m should be expected for $D \cong D_c$. For molecules with $D \lesssim 2.0$ debye, the Born approximation tends to underestimate σ_m, probably because it makes no allowance for distortion of the scattered wave (Garrett, 1972).

For strongly polar molecules, the Born approximation overestimates σ_m due to the neglect of "back-coupling" and the consequent violation of unitarity of the S matrix. A number of theoretical studies have focused on this problem, but the most precise calculations are those of Collins and Norcross (1977) for $e-$LiF scattering ($D \cong 6.5$ debye for LiF). The purpose of this study was to investigate the influence of short-range interactions on the differential and momentum-transfer cross sections. The calculations were carried out in the body-frame (fixed-nuclei, rotational impulse approximation) for the lower partial waves and in the laboratory frame (full rotational close coupling) for large partial waves. Their results are compared in Fig. 5.18 for several models: SE (static-exchange, free-electron-gas approximation), S (static), DCO (cut-off dipole models), BI (Born approximation, dipole only), BII (unitarized Born approximation, dipole only). The resonance in σ_m at $E \cong 2$ eV is present only in the SE calculation. Away from the resonance BII agrees fairly well with the other models. While this agreement may be fortuitous, BII is very simple to apply and its possible general applicability to such systems deserves further study. A recent measurement of the differential cross section at 5.44 eV by Vuskovic et al. (1977) agrees well with all the models for $\theta \lesssim 60°$ but tends to favor the SE results at larger angles. Collins and Norcross have made similar studies for NaF, NaCl, and LiCl finding resonances in all but the last case.

The strong long-range interaction which characterizes collisions of electrons with polar molecules lends itself nicely to a semiclassical treatment. Smith and Miller (1977) have formulated the semiclassical treatment of electron-molecule scattering and Smith et al. (1975) and Hickman and Smith (1977) have applied the theory to electron scattering from polar molecules. The method is a semiclassical perturbation scattering (SPS) theory based on a simple semiclassical representation of the S matrix for a wide class of e-molecule interaction potentials. It reduces to Born in the appropriate "weak-coupling" limit and is felt to be superior to Born in the

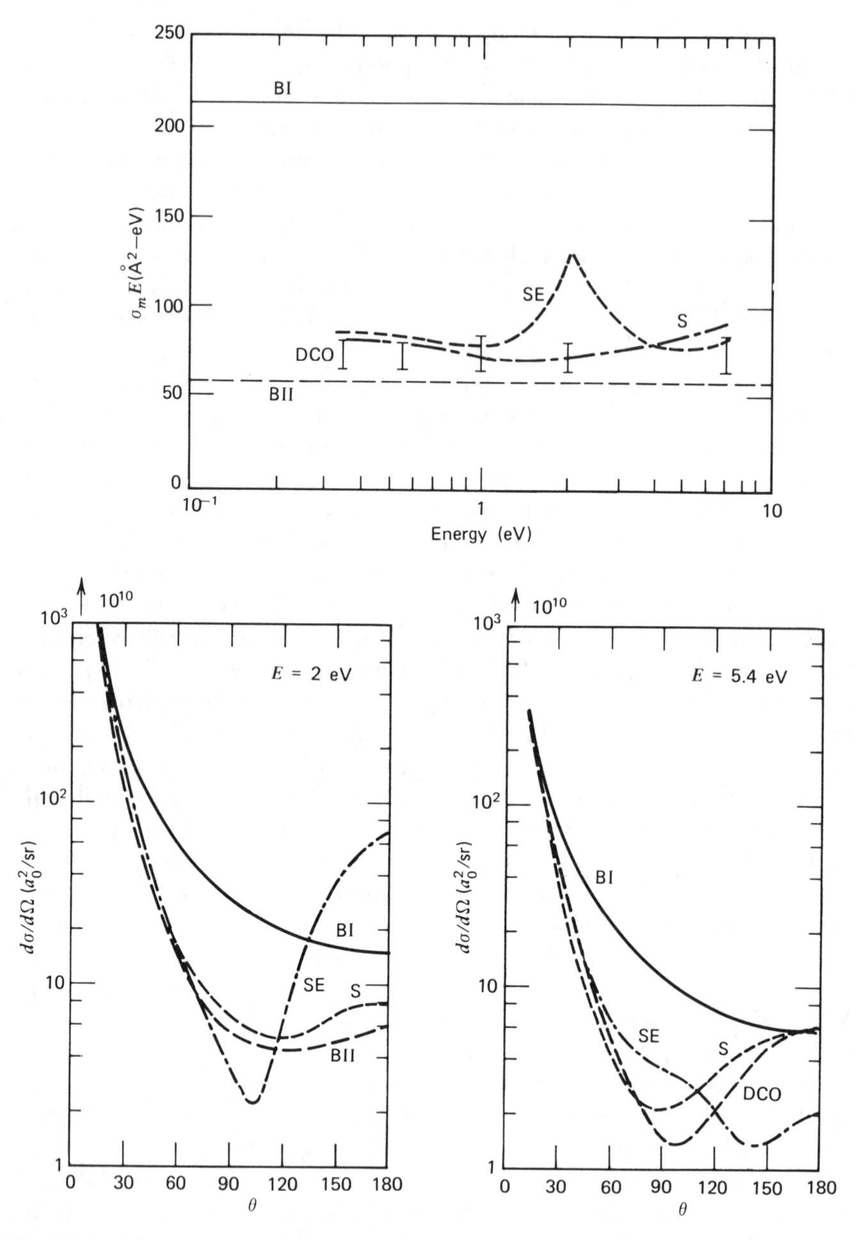

Fig. 5.18 (a) Theoretical momentum-transfer cross sections for electron scattering from LiF. (b) Differential cross sections at 2.0 and 5.44 eV. BI and BII are first Born and unitarized Born approximations; SE is static-exchange; S is static; and DCO is dipole cut-off model potential. (From Collins and Norcross, 1978.)

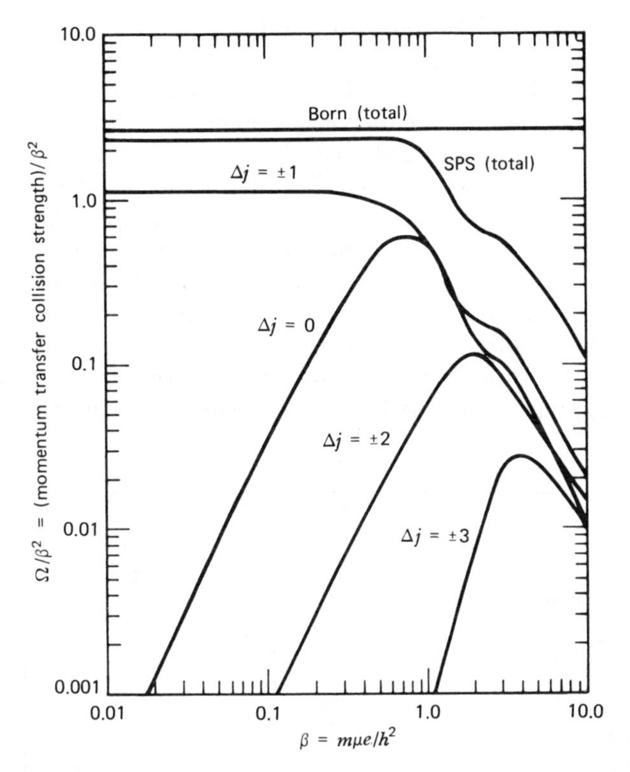

Fig. 5.19 Comparison of "reduced" theoretical momentum-transfer collision strengths in the Born and semiclassical scattering (SPS) approximations for a range of "reduced" dipole moments. SPS collision strengths for elastic scattering ($\Delta j = 0$), rotational-excitation ($\Delta j = \pm 1, \pm 2, \pm 3$), and totals are shown. (From Hickman and Smith, 1977.)

strong-coupling regime). Figure 5.19 (from Hickman and Smith, 1977) illustrates the nature of their results for the momentum transfer cross section at $E \cong 0.03$ eV. Aside from constant factors, this is a plot of rotational-excitation cross section divided by D^2 versus D, where D is the dipole moment of the molecule. The SPS theory reduces to Born for small D and is clearly superior for large D. In the case of LiF, the SPS value for σ_m is close to the BII result calculated by Collins and Norcross. While further tests against more precise calculations are desirable, SPS has real promise as a simple theoretical method, applicable in principle to a wide class of polar molecules under conditions where short-range interactions are not important.

3.9 Polar Molecules: Vibrational Excitation

In the case of vibrational excitation of polar molecules it was certainly reasonable to assume that the threshold region would be controlled by the long-range dipole coupling and that the Born approximation would be valid. Figure 5.20 shows the experimental results of Rohr and Linder (1976) for HCl. A single narrow resonance is apparent just above threshold for each transition. Similar threshold resonances have been observed for a number of other molecules as well. Gianturco and Rahman (1977) have shown formally that a "virtual state" enhancement of the final-state function could produce such a resonance.

Independently, Dubé and Herzenberg (1977) have carried out a model calculation appropriate to the HCl system which does indeed yield sharp

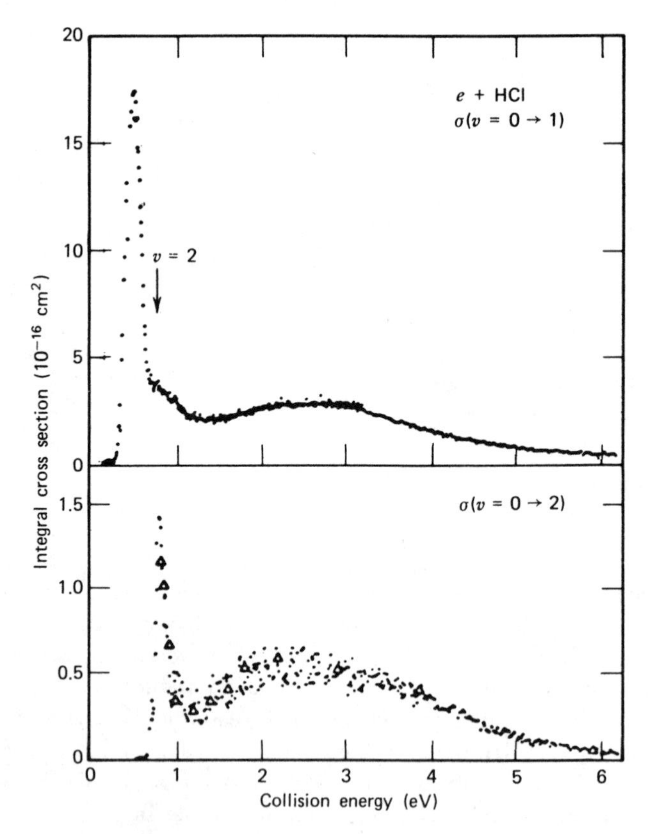

Fig. 5.20 Integral cross sections for vibrational excitation in HCl. (From Rohr and Linder, 1976, Linder, 1977.)

threshold resonances in the vibrational excitation cross sections due to a "virtual state" mechanism.

Taylor, Goldstein, and Segal (1977) have found relatively stable roots of HCl^- lying just above and essentially parallel with the ground-state HCl energy curve over a wide range in R. Such a quasi-stable state, either resonance or "virtual state" could give rise to the type of threshold resonances that have been observed.

The vibrational threshold resonances may be general phenomena, not specific to polar molecules. The Kaiserslautern group has seen such resonances in nonpolar molecules. At Rice we are working on models in hope of providing a general classification of these features in terms of the chemical makeup of the target molecule.

In conclusion, let us emphasise the importance of an even closer working relationship between theory and experiment than we have had in the past. The model studies are particularly valuable in the interpretation of scattering processes involving complex molecules. The near-*ab-initio* calculations should be continued by those best qualified to do, and they should be applied to processes and targets simple enough to permit precise quantitative analysis. The model studies should be calibrated against these accurate results as well as the results of careful experimental measurement at every opportunity. Regardless of the approach the objectives are the same—to provide fundamental understanding of the transient e-molecule state development and to obtain new information about the structure of molecules.

REFERENCES

Ardill, R. W. B. and W. D. Davison, 1968, *Proc. Roy. Soc.*, **A304**, 465.

Baille, P. and J. W. Darewych, 1977, *Proc. 10th ICPEAC*, p. 502.

Birtwistle, D. T. and A. Herzenberg, 1971, *J. Phys.*, **B4**, 53.

Brüche, E., 1927a, *Ann. Phys.*, **82**, 912.

Brüche, E., 1927b, *Ann. Phys. (Leipz)*, **83**, 1065; see also R. B. Brode, 1933, *Rev. Mod. Phys.*, **5**, 257.

Buckley, B. D. and P. G. Burke, 1977, *J. Phys.*, **B10**, 725.

Burke, P. G. and N. Chandra, 1972, *J. Phys.*, **B5**, 1696.

Burke, P. G. and A. L. Sinfailam, 1970, *J. Phys.*, **B3**, 641.

Cadez, I., F. Gresteau, M. Tronc, and T. I. Hall, 1977, *Proc. 10th ICPEAC*, p. 146; see also *J. Phys.*, **B7**, L132 (1974).

Chandra, N. and A. Temkin, 1976, *Phys. Rev.*, **A13**, 188.

Chang, E. S. and U. Fano, 1972, *Phys. Rev.*, **A6**, 173.

Chang, E. S. and A. Temkin, 1970, *Phys. Rev. Lett.*, **23**, 399.

Chase, D. M., 1956, *Phys. Rev.*, **104**, 838.
Chen, J. C. Y., 1964, *J. Chem. Phys.*, **40**, 3513.
Chen, J. C. Y., 1969, *Adv. Rad. Chem.*, **1**, 245.
Claydon, C. R., G. A. Segal, and H. S. Taylor, 1970, *J. Chem. Phys.*, **52**, 3387.
Collins, L. A. and D. W. Norcross, 1977, *Phys. Rev. Lett.*, **38**, 1208.
Crees, M. A. and D. L. Moores, 1975, *J. Phys.*, **B8**, L195; see also **10**, L225 (1977).
Crompton, R. W., D. K. Gibson, and A. I. McIntosh, 1969, *Austr. J. Phys.*, **22**, 715.
Dalgarno, A. and R. J. Moffett, 1963, *Proc. Nat. Acad. Sic. India*, **A33**, 511.
Domeke, W. and L. S. Cederbaum, 1977, *Phys. Rev.*, **A16**, 1465.
Dubé, L. and A. Herzenberg, 1977, *Phys. Rev. Lett.*, **38**, 820.
Ehrhardt, H. and F. Linder, 1968, *Phys. Rev. Lett.*, **21**, 419.
Ehrhardt, H. and K. Willman, 1967, *Z. Phys.*, **204**, 462.
Fisk, J. B., 1936, *Phys. Rev.*, **49**, 167.
Fliflet, A. W., D. A. Levin, M. Ma, and V. McKoy, 1977, to be published.
Garrett, W. R., 1972, *Mol. Phys.*, **24**, 465.
Gerjuoy, E. and S. Stein, 1955, *Phys. Rev.*, **97**, 1671; **98**, 1848.
Gianturco, F. A. and N. K. Rahman, 1977, *Chem. Phys. Lett.*, **48**, 380.
Golden, D. E., 1966, *Phys. Rev. Lett.*, **17**, 847.
Golden, D. E., H. W. Bandel, and J. A. Salerno, 1966, *Phys. Rev.*, **146**, 40.
Golden, D. E., N. F. Lane, A. Temkin, and E. Gerjuoy, 1971, *Rev. Mod. Phys.*, **43**, 642.
Hara, S., 1967, *J. Phys. Soc. Japan*, **22**, 710.
Hara, S., 1969a, *J. Phys. Soc. Japan*, **27**, 1009.
Hara, S., 1969b, *J. Phys. Soc. Japan*, **27**, 1592.
Hasted, J. B. and A. M. Awan, 1969, *J. Phys.*, **B2**, 367.
Henry, R. J. W., 1970, *Phys. Rev.*, **A2**, 1349.
Henry, R. J. W. and N. F. Lane, 1972, *Phys. Rev.*, **A5**, 296.
Herzenberg, A. and F. Mandl, 1962, *Proc. Roy. Soc.*, **A270**, 48.
Hickman, A. P. and F. T. Smith, 1977, *Proc. 10th ICPEAC*, p. 492.
Itikawa, Y., 1971, *Phys. Rev.*, **A3**, 831.
Kaldor, U. and A. Klonover, *Proc. 10th ICPEAC*, p. 488.
Krauss, M. and D. Neumann, 1972, *Chem. Phys. Lett.*, **14**, 26.
Lane, N. F. and S. Geltman, 1967, *Phys. Rev.*, **160**, 53.
Linder, F., 1969, *Proc. 6th ICPEAC*, p. 141.
Linder, F., 1977, private communication.
Lowke, J. J., A. V. Phelps, and B. W. Irwin, 1973, *J. Appl. Phys.*, **44**, 4664.
Massey, H. S. W., 1932, *Proc. Camb. Phil. Soc.*, **28**, 99.
Massey, H. S. W., 1969, *Electronic and Ionic Impact Phenomena*, vol. II, Oxford Univ. Press, London.
Massey, H. S. W. and R. O. Ridley, 1956, *Proc. Phys. Soc.*, **A69**, 659.
Morrison, M. A. and N. F. Lane, 1977, *Phys. Rev.*, **A16**, 975.
Morrison, M. A., N. F. Lane, and L. A. Collins, 1977, *Phys. Rev.*, **A15**, 2186.
Morrison, M. A. and B. I. Schneider, 1977. *Phys. Rev.*, **A16**, 1003.

Nagahara, S., 1954, *J. Phys. Soc. Japan*, **9**, 52.

Pacansky, J., U. Wahlgren, and P. S. Bagus, 1975, *J. Chem. Phys.*, **62**, 2740.

Ramsauer, C. and R. Kollath, 1927, *Ann. Phys. (Leipz.)*, **83**, 1129; 1932, ibid., **15**, 485.

Ramsauer, C. and R. Kollath, 1929, *Ann. Phys.*, **4**, 91.

Rescigno, T. N., C. W. McCurdy, and V. McKoy, 1974, *Chem. Phys. Lett.*, **27**, 401; 1975, *Phys. Rev.*, **11**, 825.

Rescigno, T. N., C. Bender, C. W. McCurdy, and V. McKoy, 1976, *J. Phys.*, **B9**, 2141.

Riley, M. E. and D. G. Truhlar, 1975, *J. Chem. Phys.*, **63**, 2182.

Rohr, K. and F. Linder, 1976, *J. Phys.*, **B9**, 2521.

Rohr, K. and F. Linder, 1977, *J. Phys.*, **B** to be published.

Schneider, B. I., 1975, *Chem. Phys. Lett.*, **31**, 237; *Phys. Rev.* **11**, 1957.

Schneider, B. I., 1976, *Phys. Rev.*, **A14**, 1923.

Schneider, B. I. and P. J. Hay, 1976, *Phys. Rev.*, **A13**, 2049.

Schulz, G. J., 1962, *Phys. Rev.*, **125**, 229.

Schulz, G. J., 1964, *Phys. Rev.*, **135**, A988.

Shugard, M. and A. U. Hazi, 1975, *Phys. Rev.*, **A12**, 1895.

Smith, F. T., D. L. Huestis, D. Mukherjee, and, H. W. Miller, 1975, *Phys. Rev. Lett.*, **35**, 1073.

Smith, F. T. and W. H. Miller, 1977, *Proc. 10th ICPEAC*, p. 484.

Takayanagi, K. and Y. Itikawa, 1970, *Adv. Atom. Mol. Phys.*, **6**, 105.

Taylor, H. S., 1970, *Adv. Chem. Phys.*, **17**, 91.

Taylor, H. S., E. Goldstein, and G. A. Segal, 1977, *J. Phys.*, **B10**, 2253.

Tully, J. C. and R. S. Berry, 1969, *J. Chem. Phys.* **51**, 2056.

Vuskovic, L., S. Strivastava, and S. Trajmar, 1977, *Proc. 10th ICPEAC*, p. 658.

Wilkins, R. L. and H. S. Taylor, 1967, *J. Chem. Phys.*, **47**, 3532 (because of an error in numerical procedure, these results are only qualitatively significant).

CHAPTER 6

Problems and Possibilities in the Study of Electron-Molecule Collisions

HARRIE S. W. MASSEY

Department of Physics and Astronomy
University College London
London, England

In introducing this discussion there is no need to emphasize that, because of its concern with molecules, the subject we are concerned with is of as much interest to chemists as to physicists. These are of course differences between the aims and methods of chemists and physicists that enable their contributions to be complementary. The chemist normally works with very complex systems and of necessity must be more empirical in his approach. The physicist on the other hand aims at a deeper understanding of the underlying factors determining a particular process and therefore tends to confine his attention to relatively simple systems. Electron-molecule scattering already drags him into more complex situations than he is accustomed to, but his basic aim of deeper understanding, though hard to fulfill, remains valuable even though it must be diluted somewhat.

Because of its interdisciplinary character it is appropriate to begin by discussing *e*-molecule scattering, paying particular attention to its relation to the deeper understanding of molecule structure. We then consider collisions in which the electron is captured, either on impact with a neutral molecule to form a negative ion or with a molecular positive ion leading to recombination.

1 ELECTRON SCATTERING

1.1 Elastic Scattering

Can we get any real information about the structure of a molecule from the elastic scattering of slow electrons by it as distinct from being able to predict the rates of the scattering processes? This is an uncertain matter. There is a great difference between finding out something about the structure from the scattering of slow electrons and the sort of information obtainable about that structure from x-ray diffraction. In the latter case you are dealing with the unmodified molecule. X-rays do not effectively modify the molecular structure during the scattering process although there is an intimate interaction. It is uncertain as to how much information one can get about the structure of the target, in the absence of the electron, from studying what happens when the electron is scattered by it. The question can only be answered by much more intensive study, both experimental and theoretical, of the elastic scattering by molecules.

On the experimental side this requires accurate absolute measurements of cross sections and it is only fair to say that atomic collision physics is not yet a precision branch of physics. It is somewhat humiliating when the nuclear physicists are able to make measurements of much higher accuracy under many circumstances similar to those that are to be found in the molecular area. It is important to compare the accuracy of absolute measurements of molecular scattering cross sections so that, for example, a phase analysis may be made in the case of symmetrical diatomic molucles.

Next let us turn our attention to accurate calculations of both *direct* and resonance contributions to scattering cross sections. "Direct" is italicized because there has been so much interest in the resonance contributions that one wonders sometimes whether it has been forgotten that we also need to know quite accurately what happens in direct scattering. This is particularly true in the case of N_2, although there are other cases. What needs to be emphasized here is that much more work needs to be done in accurate calculations using methods in addition to those commonly employed for resonance scattering.

What are the prospects of handling scattering by polyatomic molecules theoretically? They may not be good, but we should not be too pessimistic at the outset. If one is really to make a contribution to understanding molecular structure, in general, is it not too restrictive to consider only diatomic molecules? Various attempts are being made to handle the polyatomic molecule problem and we have already obtained quite a lot of valuable and important information, but we need to be patient—success will not come quickly. Nevertheless modern computers are now so powerful that we can expect progress to be made. It is important, however, that computers should be used to assist understanding rather than merely to provide "brute force" solutions of particular problems, important as such solutions may be for special applications.

1.2 Coincidence Measurements for (e, 2e) Reactions in which the Emitted Electrons have Comparable Energies

The development of coincidence counting methods in atomic physics has made available a valuable new tool for investigating molecular structure. The technique consists in observing the separate moments of the two electrons resulting from an ionizing collision of an electron with a molecule under suitable geometrical and energetic conditions. If the energies of the primary and both secondary electrons are large compared with the ionization energy, the collision is effectively one of the incident electron with a free molecular electron whose momentum is determined by the molecular momentum eigenfunction. If the measurements are made in which the secondary electrons have equal energies and are emitted in directions making equal angles with that of the primary, but are not coplanar, the variation of probability with the azimuthal angle between the planes of emission gives directly the square of the momentum eigenfunction averaged over initial vibrational and rotational states. This technique has already produced interesting results for some molecules and is clearly one of great potential for investigating molecular structure.

1.3 Molecular Structure from Ion-Cracking Fraction Studies

A third subject can be introduced by the question—can any information about molecular structure be devised from ion-cracking fraction studies? A lot of measurements have been made in the past of the fraction of different ions produced from a particular molecular gas when it is bombarded by electrons of different energy. These have been made for various reasons connected with the oil industry and other applications in one way or another. They have produced a vast amount of data, but little attempt has been made to understand them. The data have been collected for the purpose of making some direct application and then forgotten. There is a good deal of information here that may be of some value.

2 INELASTIC COLLISIONS LEADING TO MOLECULAR DISSOCIATION

The production of molecular dissociation into neutral fragments by electron impact is often of importance in practice as well as being of considerable basic interest. The experimental study of these collisions is relatively difficult because of the need to detect the neutral fragments, especially if these fragments are in their lowest electronic states. Even the classical example of the dissociation of H_2 into two normal H atoms has not yet been thoroughly investigated either experimentally or theoretically. There is scope here for much more detailed work taking advantage of modern techniques.

3 AUTODETACHMENT

What is the definition of a resonance? It is surprising in a field which has been studied as much as electron-molecule collisions that the definition of resonance is still unclear, but it is particularly important when discussing attachment phenomena that what we really mean by a resonance be resolved.

The special molecular effects that are associated with resonance effects need to be thoroughly understood. These include such things as autodetachment through the coupling of electronic with vibrational and rotational motion; autodetaching Rydberg states; and the possible occurrence of steric effects for polyatomic molecules.

We can also expect more important results on the dependence of attachment cross sections on the initial state of internal nuclear motion. Information of this kind is not only important for its own sake, but for the evidence it provides about the slopes of potential energy curves and for polyatomic molecules, the curvature of potential energy surfaces.

With the wealth of new techniques available we should now be able to analyse in some detail what happens when a highly symmetrical molecule like SF_6 forms SF_6^- by attachment of electrons of almost zero energy. Can we yet follow through the internal energy transfer processes that lead to a long lifetime for the complex formed initially by the electron capture? Comparison with other molecules such as SeF_6, TeF_6 and UF_6 might be helpful here as well as a strong input from chemistry.

4 COLLISIONS OF ELECTRONS AND MOLECULAR IONS

A very important process of recombination is that between an electron and a molecular positive ion in which the energy released by capture of the electron is used up in dissociating the neutral molecule so formed viz.

$$AB^+ + e \rightarrow A + B$$

where the neutral products may or may not be excited. Such dissociative recombination is not only of major importance in planetary ionospheres, including that of the earth, but also in determining the nature of the molecules in interstellar space. For an understanding of modes of formation of these molecules the rates of dissociative recombination to ions such as CH^+ need to be known with reasonable accuracy. As these ions are in their lowest vibrational state under interstellar conditions and as the recombination rate is sensitive to the initial vibrational state, it is essential that laboratory measurements made refer to ions without vibrational excitation. There is still a need for such measurements for ions of astrophysical interest as well as for continued theoretical research on the mechanism of recombination to polyatomic as well as diatomic ions.

Author Index

Subject Index